HERETICS AND HEROES

IN SCIENCE

by Frederick Goldman

1stBooks – rev. 07/11/01

TABLE OF CONTENTS

1. The Crucible of GIORDANO BRUNO 1

2. HELEN CALDICOTT   Blew the whistle on Nuclear proliferation............. 8

3. BENJAMIN CARSON, M.D.   From ghetto to Worldwide fame 19

4. RACHEL CARSON   Launched a revolution.................................... 29

5. MARIE SKLODOWSKA CURIE   Opened the door to Science for women. 39

6. But for fate, "Origin of the Species" might not have been authored by CHARLES DARWIN 49

7. DOROTHEA DIX   Advocate for those who had no other advocate 60

8. CHARLES DREW   Pioneered the struggle to emancipate Black physicians.. 73

9. RICHARD FEYNMAN   The twelfth man blew the whistle.................... 87

10. ROSALIND FRANKLIN   discovered the structure of DNA 96

11. GALILEO GALILEI   Hero or Craven? 109

12. JOHN HARRISON   Unschooled, he succeeded where Galileo and Newton failed. 118

13. WERNER von HEISENBERG   Still an open question: did he deny the A-bomb to Hitler? 127

14. SIR WILLIAM LAWRENCE   Paved the way for Charles Darwin.......... 138

15. LISE MEITNER   emerges from shadow of the A-bomb..................... 145

16. FLORENCE NIGHTINGALE   Triumphed over a gaggle of goliaths....... 159

17. HAROLD RIDLEY   Survived ridicule to achieve fame......................... 175

18. IGNATZ SEMMELWEIS   Reviled in life, revered in death. 181

19. NIKOLA TESLA   Once more famous than Tom Edison 186

20. THE WRIGHT BROTHERS   Started late but won the race................... 194

PREFACE

Many scientists pursue independent and unorthodox ideas – and achieve important success – despite resistance and hostility.

Why should unconventional ideas have that effect? Being independent is not popular. In Japan, parents warn their young, "The nail that sticks up gets hammered down." American parents give similar guidance: "To get along, go along."

Conformity is a natural instinct among humans, as among animals. But unlike animals, humans are gifted with intelligence beyond the impulse of instinct. We learn from the past. We store up knowledge.

The instinct to conform collides with the urge toward independence and innovation – the two represent a push-pull in human behavior. This sometimes produces conflict when those in power see "change" as dangerous to themselves, their professions, their religion, or the conventions of their society. Many important discoveries and achievements – in every field of human endeavor – were at first opposed by those in authority. This struggle is particularly significant in the sciences. The essays herein collected profile examples of progress that resulted from successful contest with resistance.

Some individuals draw from their treasury of experience and knowledge to work on unmet needs; or to develop new inventions, new art forms – new ideas to improve the quality of life for mankind.

These essays celebrate but a few who advanced the cause of civilization; their number is great. Some of the individuals here nominated were famous in their time; some were not. All did important things. I provide only highlights of their lives – a tiny fraction of the information available. I urge further exploration.

The Crucible of
GIORDANO BRUNO

The defrocked priest was gagged and bound to a stake amid drumbeats, 400 years ago. A torch was thrown at his feet into the pyre of straw. As he was enveloped by flames, a messenger hastened to the Vatican, nearby, where Pope Clement, having prescribed that punishment, awaited. Thus ended, at 52, on February 17, 1600, in Rome's Campo de' Fiori, the extraordinary life of Giordano Bruno. The Encyclopedia Britannica identifies him as "one of the important figures in the history of Western thought, a precursor of modern civilization;" the "Dictionary of Scientific Biography" opines he represented a "great turning point in human history." He profoundly influenced Descartes, Spinoza, and Liebnitz.

In the papal verdict that condemned Bruno, eight specific heresies were listed – all of them extracted from his writings. Such as his stinging rebuke for the Inquisition policy of forced conversions. "The procedure that the Church uses today is not that which the apostles used, for they converted the people with ... the example of a good life, but now whoever does not wish to be a Catholic must endure punishment and pain, for force is used and not love; ... the Catholic religion has need of great reform ..." He assailed fellow clerics as "senseless and foolish idolaters ... who seek the divinity of which they know nothing, in the excrements of dead and inanimate things." Even Martin Luther had not used such harsh language.

Born at Nola, Italy, near Naples, and baptized Filippo, Bruno took the name Giordano when he entered the Convent of San Domenico Maggiore in his teens. There he mastered the techniques of memory retention and Aristotelian philosophy for which the Dominicans were noted. Ordained a priest at 19, he knew, word for word, the teachings of Thomas Aquinas, who had lived and worked at that institution. His piety and phenomenal memory were reported to Pope Pius, who sent a carriage from Rome so the young man might perform for eminences in the Vatican.

He was of independent mind. An adventurous reader, he found in the convent library "forbidden" commentaries on church history by the humanist philosopher Erasmus which caused his first brush with Authority. He discussed with colleagues the Arian Catholic teaching that denied the divinity of Jesus, doubted the historical veracity of immaculate conception, and questioned the doctrine of transubstantiation. They denounced him to the provincial father. He was excommunicated and a trial for heresy was planned. He fled the convent in 1576. The Inquisition traced him to Rome, and a second excommunication was levied. He shed his Dominican robes and, as a layman, disappeared. In 1578, he reached Geneva, then a center of Reformation fervor. He renounced membership in the Dominican Order and embraced Calvinism.

Which proved no less intolerant of dissent in dogma when he disagreed with Luther's doctrine that salvation could be won through faith alone. He argued that good works were necessary, too – as demanded in the Gospel of St. James. Bruno was again excommunicated, but punishment was withheld on condition he leave Geneva promptly.

He found refuge in Toulouse, France. There he lectured on philosophy and memory-enhancement. His reputation reached the ears of King Henri. Summoned to Paris in 1581, Bruno was salaried as a "royal lecturer," and enjoyed two years of security. During this time, he commenced a prolific writing career: allegories and Socratic parodies; philosophic meditations; poems, even a play.

He also studied the principles of a "heliocentric" universe, formulated 30 years before by Nicholaus Copernicus. He became an adherent of the Polish friar/physician/astronomer, who had written the "Sun ... governs the surrounding family of the stars" displacing the earth as the center of the universe. (A prudent man, Copernicus had, for years, shelved his book, "De revolutionibus orbium caelestium," until he was 70 and nearing death. He then sent a copy to the pope, with a wry preface in which he expressed fear of misunderstanding by ignorant readers. He meant, of course, all who clung to the Ptolemaic geocentric dictum – and church dogma – that the earth was axis of the solar system. Copernicus was excommunicated, his book banned.)

In the spring of 1583, Bruno was sent by King Henri to London, where he became a counselor to the French ambassador, Michel de Castenau. At Queen Elizabeth's court, he early befriended the influential Sir Philip Sidney and the Earl of Leicester, Robert Dudley. Invited by them to deliver a series of lectures at Oxford that summer, he opened his first address by assuring the university's vice-chancellor and fellows that he was "a stranger nowhere save amongst the barbarous and ignoble, the waker of sleeping souls; tamer of presumptuous and recalcitrant ignorance ... who does not choose ... the mitred head more than the crowned head ... the cowled man more than the man without a cowl, but him who is more civilized, the more loyal, the more useful ... (pursuing) the culture of the mind and soul ... (though) hated by the propagators of foolishness and hypocrites ..."

Many in Bruno's audience had witnessed, in 1550, the burning of books and manuscripts when the King of England's commissioners ransacked Oxford libraries – particularly books containing mathematical diagrams thought to be "Popish" by the newly-Protestant regime. Bruno went on to expound Copernicus' sun-centered theory. He challenged: "Who, indeed in this most magnificent temple, would put the light in another place ... some people call it the lamp of the world, others its mind, others its ruler ... the visible God ... the All-Seeing."

But Bruno's basic piety launched a more profound vision: "the excellence of God (has) magnified ... the greatness of his kingdom (and is) glorified not in one, but in countless suns; not in a single earth, but in a thousand ... in an infinity of worlds." Which, he proposed, populated an infinite number of solar systems similar to ours. In essence, his was a monistic conviction — the basic unity of all substances, spirit, form and matter. He said God and nature could not possibly be separate and distinct entities, as church doctrine insisted, still in thrall to the authority of Aristotle. Two thousand years after the Greek sage had died, his extant writings were still taught, though already some of his axioms were known to be false. At Oxford, Aristotle was still sacred. An ordinance enjoined faculty "to follow Aristotle faithfully" or "be liable to a fine of five shillings on every point of divergence."

Warming to his polemic, Bruno then attacked orthodoxy per se. He declared that the Bible was authoritative only in moral issues; Nature and science followed laws of their own. Religion was simply a means to instruct and govern the unlettered and ignorant – and was often perpetuated by clerics in naked self-interest. He dismissed the mysteries of faith; suggested that Old Testament "history" was no more reliable than Greek myths and labeled New Testament "miracles" as nothing but magic tricks. Prayer was a waste of time. He indicated men of the cloth as responsible for transforming Jesus' simple message of love into hair-splitting casuistry. He termed them "pedants" who were "avaricious and dissolute." Those who trafficked in superstition he likened to an ass which had trespassed into churches, courts of law, even colleges:

> "... no one works for them, and they work for no one ... yet they live on the work of ... those who have temples, chapels, inns, hospitals, colleges and universities; ... necessary to the commonwealth, skilled in the speculative sciences, careful of morality, solicitous for increasing zeal and care for helping one another and maintaining society ..."

For good measure, he went on to castigate the rigid English social system and sneered at "crude" English manners. It might be assumed that the Oxford dons choked when they heard these condemnations. He later described their "rough and rudely" reaction, in a biographical essay. The "series" of lectures was abruptly terminated when Queen Elizabeth was told that Bruno described Christianity as irrational; revelation was a fiction; the virginity of Mary was untenable; the Mass meaningless.

He became a pariah in British society, a subversive, almost a fugitive. He took refuge in the French Embassy, which he didn't dare to leave for weeks, when mobs gathered outside. During this time, he composed essays that later became famous, reiterating that every material substance is a manifestation of a godhead. Deeply influenced by Egyptian religion, he nominated the sun for that role.

He left England as soon as he could. The political weathervane in Paris had, in the interim, veered away from religious liberalism, and toward the intransigence that had inflamed the papal-approved St. Bartholomew's Day Massacre of barely a decade before. King Henri had, since, protected the protestant Huguenots – but it was now expedient to retreat from being their defender. The right-wing Catholic-League, bankrolled by newly-wealthy Spain, replaced the king's milder-mannered counselors.

Bruno left Paris and trailed from one university town to another: Marburg, Frankfurt, Prague. He subsisted by translating, proof-reading, lecturing. And he refined his own religious credo:

> "God as a whole is in all things … in a crocus, a daffodil, a sunflower … divinity descends … as it communicates itself to nature, so there is an ascent made to divinity through nature.
>
> True philosophy is music, poetry, or painting; true painting is poetry, music or philosophy; true poetry or music is divine sophia and painting."

He opined that the earth is ordained to "renew" itself … "for it … cannot be eternal … substances which cannot be everlasting … change themselves. Therefore, since death and dissolution are unfitted to … this globe, this star, … and complete annihilation is impossible to all nature … (as) nothing is of itself eternal, save the substances and material of which it is made, and this is in constant mutation."

Bruno was the first to say "There is no absolute up or down … no absolute position in space; but the position of a body is relative to that of other bodies. Everywhere there is incessant relative change in position throughout the universe, and the observer is always at the center …"

All human perceptions hence are relative to the viewer's position in space and time. (Three hundred years later, Einstein built upon this principle.) This argument, of course, rebutted Rome's claim of omniscience and absolute authority. Audacious concepts like these spread his fame. The rector of the university in Wittenberg (Luther's base) appointed the celebrated wanderer to a faculty post which commenced the happiest period in the man's life. (He memorialized it in print.) Flushed with success, he accepted invitations to lecture at German, Swiss, and Czech cities, Catholic and Protestant. In addition to amplifying his criticisms of mainstream religious practices, he urged an ecumenical armistice, and called for tolerance between all sects. Stressing the soul is immortal – a belief shared by Christians of all denominations – he proposed discussions to heal the schism among the churches.   Though recognizing the need for sweeping reforms in Roman rule, he renewed his vows in rejoining his original commitment: the "Catholic religion pleased him more than any other."

In this euphoric state, he received an invitation from Giovanni Moncenigo, patrician in Venice, to teach his techniques of memory enhancement there.

He had not been in Italy since excommunication by Catholics and rejection by Protestants.   Why should he even contemplate taking the "fatal step of returning" to hostile territory, asks Francis Yates, Bruno's distinguished biographer.   She offers this answer: "People like Giordano Bruno are immunized from a sense of danger by their sense of mission ... euphoria bordering on insanity in which they constantly live."

Though Roman Catholic, Venice was cosmopolitan, almost completely independent of Vatican influence.   But first, Bruno stopped at Padua, within the Vatican Republic.   The faculty chair in mathematics at the famed university was vacant.   He applied.   While waiting for a decision, Bruno earned his daily bread tutoring students on the Dominican arts of mnemonics.   The professional appointment went not to him, but to a younger man, also reputed to be brilliant, but politically uncontroversial: Galileo Galilei.   (The latter was also an adherent of the Copernican theory, but as he admitted in a letter to Johannes Kepler, his fealty was "silent."

Bruno reached Venice in March of 1591.   Another disappointment followed – rejection by the "pupil" (and host) whose memory did not expand appreciably under instruction.   Moncenigo (was he an agent?) denounced Bruno to the Inquisition.   His printed works were used in an indictment of heresy. Interrogations began May 25, 1592.   A witness quoted him as saying that the Roman Church had "stolen" the sign of the cross from Egyptian sculptures of the goddess Isis.   Another prisoner related Bruno's belief that Moses was a seer whose powers had been acquired from Egyptian priests – and used them to confound Pharaoh's wise men.   Christ was likened in magic powers to Moses.

The Inquisitors reviewed Bruno's often-printed belief in a sun-centered universe which was infinitely large and consisted of an infinite number of worlds.   Divine power could not create a finite world, he had written.   Nature is the visible shadow of God, ineffable and inexplicable, he repeated under interrogation.   His captors were stunned to hear that anchored-in-nature Egyptian cults, suppressed by Christianity, should be re-embraced within the Roman Catholic faith.

Venice was spared the responsibility of rendering a verdict.   He was released on an extradition order from Rome.   There, in the catacombs of the Holy Office, he faced the formidable Cardinal Robert Bellarmine, theological advisor to the pope.

What were the basics of Christian "theology" 400 years ago? From the time of St. Paul, Christianity had spread most rapidly among pagan societies in the Roman empire – and absorbed some of their culture. Fables, superstitions, and fantasies had dominated through the dark centuries. By adapting to them – in some cases absorbing them – the Church had subsumed some barbarian beliefs and preserved remnants of western civilization. Hanging in the "Borgia apartment" in the Vatican, a painting in the Renaissance fashion depicts the Egyptian goddess Isis attended by Moses and an early Christian saint. Another, by the same artist, represents the Egyptian bull-god Apis, being worshipped by Christian priests. The sarcophagus of a sage in the Christian catacombs at Beth She'arim in Israel is embellished with the head of Zeus. Pope Clement himself was a firm believer in astrology, and had a chamber in the Vatican reserved for occult incantations.

Even more than today, ordinary people were interested in esoterica – cults, magic, sighting, visions, alchemy, local and regional saints and superstitions, angels, demons. The invention of printing in the previous century had enabled imaginative writers to publish "eye-witness" reports of occult events, complete with mysterious signs and symbols, formulae, even maps and diagrams.

Cardinal Bellarmine recognized the ex-priest's deep piety. His hostility to Aristotle was far from unique: by the late 16th Century, many of the Greek's so-called "axioms" had already become obsolete. Cardinal Bellarmine read Bruno's books and the transcript of his Venetian trial. He concluded that in the spirit of the counter-Reformation, the man was a heretic, an atheist and infidel – but decided that he was not dangerous. Nor a proponent of any current aberrant digressions that undermined Roman Catholic dogma.

Its theology was in flux. St. Thomas Aquinas, Bruno's spiritual mentor, had been condemned – after his death but before he was canonized – by his own archbishop. The Franciscan Order, which had been founded on a principle of mendicancy, forbade its members to study the Thomist "Summa Theologica," wherein he strongly endorsed St. Francis' Rules of Poverty (which rules had, by Bruno's time, been determined by the Vatican to be a heresy). That Bruno disbelieved in the immaculate conception was simply following in Aquinas' footsteps. Bruno could not be faulted for his appreciation of the natural world. Nor his believing, as did Aquinas, that in matters of faith, acceptance by the believer is a matter for individual moral decision.

Later canonized as a saint, Cardinal Bellarmine himself faced a moral decision. What to do with this eccentric?

As "Giordanisti" followers north of the Alps were not actively headed by Bruno, the man posed no serious threat. Bruno had already been in prison eight years. The cardinal offered parole but commanded him to abjure in detail his outrageous statements. On his knees, Bruno denounced all that he had proselytized for years. He was released.

6

But freedom was haunted by conscience. Betrayal of all that he had believed in, was unbearable. In the hearing of others, he repudiated his retraction as not having been made in good faith. For the third time, he was jailed. The pope himself, on February 8, 1600, ordered death for the "impenitent and pertinacious heretic." Bruno wrote back defiantly "your fear in passing judgment on me is greater than mine in receiving it."

Three centuries later, a statue was erected in Rome's ancient Campo de' Fiori marketplace to memorialize where the one-time Dominican priest had been burned alive.

As an admirer since student days in Rome, the author suggests this epitaph should be added quoting Bruno's credo:

> "as a citizen and servant of the world, a child of Father Sun and Mother Earth, because he loves the world too much, must be hated, censured, persecuted and extinguished ... Time gives all and takes all away; everything changes but nothing perishes ... With this philosophy my spirit grows, my mind expands ... I await daybreak. Rejoice therefore ... and return love for love."

HELEN CALDICOTT
Blew the whistle on
Nuclear proliferation.

Like a meteor, she flashed brilliantly across the sky and then, like a meteor, she disappeared from sight. During her brief public career, she nudged the course of history toward sanity.

Her birth, August 7, 1939, was at the height of the worldwide Great Depression — "a brutal time in Australia's history," Helen Caldicott reflects. As a very small child, she remembers competing with rats at the town dump to harvest the berry-bushes when they were heavy-laden with fruit. They were grim and unhappy years: the oldest of three children, Helen was frequently the victim of severe beatings.

When the Japanese tide in World War II swept southward in the Pacific and overran Singapore, an invasion of Australia was expected. The children were evacuated inland, schooling interrupted.

Later, Helen's medical schooling was stop-and-go. Less than 10% of the first-year class was female, but Helen was graduated second in the class.

> "Early marriage was the norm ... a woman who was not
> married by the age of twenty-three was considered a failure and
> an old maid ... all my friends were either engaged or married."

Helen followed the norm — but became pregnant before the wedding date. Her fiancé, also a medical student was shocked: "You wouldn't do this to me just before final exams!" The baby boy came soon after the wedding; a second baby less than two years later, and a third pregnancy followed soon after. Helen was a mother of three while still in her mid-twenties — and was delighted.

Husband Bill not so. He felt trapped in fatherhood. He applied for an advanced-study fellowship in radiology at Harvard. When it was awarded, he promptly left. Helen stayed to await her third baby. She then followed Bill to Boston, got a job at Harvard's Children's Hospital, and settled down to the normal life of a doctor, wife to a doctor, a mother, a homemaker.

During the fellowship period at Harvard, Helen's mother died. The family returned to Australia, where Bill furthered his career in radiology at the Adelaide Children's Hospital. To help with family finances, Helen worked part-time at the hospital assigned to the renal clinic. She remembers "that many of the dialysis patients were infected ... and that some of the medical staff ... had developed acute hepatitis." She did, too. For five weeks she hovered between life and death. "I felt like a satellite swirling around in space with no earthly or physical contact. I was distant from the children, and I felt that my marriage to Bill was at an emotional end."

She fought depression and achieved recuperation by substituting for other doctors — wherever there was a need. At the Adelaide Hospital, she learned that children with cystic fibrosis, a fatal disease, were virtually neglected — treatment was hopeless at any event. Helen, mother of three little ones, applied for training as a pediatric intern, 80 hours a week. Upon completion of that program, she proposed that the hospital establish a clinic for cystic fibrosis-afflicted children. Her superiors assumed this was a gambit to advance her career. They turned her down: her first experience with closed doors. The rejection fueled her determination. Day and night, she poured over cystic fibrosis case histories in the hospital files. She learned that her hospital considered a 50% survival rate after four years to be the average with that malady — whereas, when she canvassed like institutions in the U.S. and Canada, actuarial data established life expectancy averaged seventeen years. The hospital management reversed its negative stand. The clinic was in due course established – the only such in Australia.

After qualifying as a pediatrician, Helen became obsessed with nuclear power as a special threat to children. For years after the Hiroshima and Nagasaki bombs, a twilight prevailed even among thinking people – what psychiatrist Robert Lipton calls "psychic numbing." Employment of nuclear weapons inflicting death and disease indiscriminately upon hundreds of thousands of civilians, was such an obscenity that it could not be faced – better to ignore it. Awareness dawned only when Russia and the U.S. started to test hydrogen bombs in 1954, sending aloft radioactive dust that floated around the globe on prevailing winds. It was dangerous to breathe. Returned to earth mixed with rain, it could also become poison, spread on crops and animals' grazing grounds.

England, Russia and the U.S. agreed that high-atmosphere testing was too dangerous. The United Nations endorsed a test-ban treaty. France, a latecomer to nuclear-testing, refused to sign the treaty. Its leader, General Charles de Gaulle, believed that national self-esteem required his country to "catch up" with the Big Three in what he called "The Nuclear Club." For years, high-atmosphere testing of French nuclear devices continued over the French-owned atoll of Muroroa in the South Pacific. Radioactive dust from those tests floated wherever winds blew; radioactive rain watered Australia and New Zealand, filling reservoirs. When drought struck, they emptied into pipelines.

Helen Caldicott had not, till then, given thought to the "passive" dangers of radioactivity. But as a pediatrician, she knew that children were far more sensitive to radioactivity than adults were. There was danger of genetic mutations as well as cancer and leukemia. Provoked into bitter criticism of the French, she wrote letters; spoke on the radio; lectured wherever an audience might be gathered, was interviewed by newspapers and magazines. Very soon, there were mass demonstrations, parades. Parents carried babies, babies carried signs: "I don't want to die of leukemia from the French tests." Protesters decided to send a delegation to Tahiti, where France's Pacific affairs had their headquarters. Helen was asked to be one of three in the group.

She asked for a 10-day leave of absence from the hospital. The Medical Superintendent, furious that his switchboard was being clogged by anti-nuclear calls, told her "If you don't stop this public campaigning, we may not be able to appoint you ... next year." She asked meekly "Has my work suffered? Have you received any complaints about my medical practice?" No, he admitted. She then asked again for the leave-of-absence. "You can't go. I won't give you permission." She purred ... "the press will want to know why I can't go." He thought for a moment. "Alright, you can go – but you're not to say anything!" A properly brought-up young lady, Helen thanked him.

The Tahiti flight was on Qantas' regular schedule. Just before departure, the airline informed the trio that the French would not grant permission to land at Tahiti. The protest team therefore flew on to Paris, and applied for an audience with the nation's president. He refused, but authorized a minor official to receive them in the Foreign Affairs Department at the Elysee Palace. Flunkies wore splendid uniforms, complete with gloves. Wearing a dress she had sewn herself, Helen iterated the dangers of radioactive fallout. The official scoffed: "Our bombs are perfectly safe." He was asked why, in that case, "don't you test your bombs in the Mediterranean?" The shocked reply: "Too many people live around the Mediterranean!"

The August 1972 junket was not a failure. It had attracted a worldwide audience and sparked worldwide indignation. At the U.N., Australia and New Zealand introduced a censure resolution. France responded by announcing a new series of tests in the atmosphere. An indictment was filed with The International Court of Justice at the Hague, in Holland, identifying the French government as a criminal defendant. As is usual when high-profile corporate lawbreakers are caught red-handed, France denied wrongdoing – but promised henceforth to conduct its tests underground. "An informed democracy behaves in a responsible fashion" is one of Helen Caldicott's favorite aphorisms.

> All the while, Helen continued her work in the hospital, by now "much more challenging than it had been when I was an undergraduate ten years earlier. Interim research had disproved many of the lessons drilled into us as medical students ... it became obvious to me that once-definitive scientific dictums changed as the research community discovers more, and I learned that it behooves scientists to retain a degree of humility.

Fulltime work at the hospital did not interfere with fulltime agitation against the nuclear threat. Australia was a major center of uranium-mining. It was providing employment for thousands of men, making profits for thousands of shareholders. Helen wrote to each and every one of Australia's 76 unions, urging attention to the issue and its relevance to the health of their members. She asked for permission to speak at union meetings. No fee was asked, no politics were involved, so permission was granted to speak at a meeting of the Adelaide Trades and Labour Council.

> " ... I'd worn a pair of black velvet slacks and an ivory-coloured satin blouse – so they might at least look at me. But ... they continued their loud conversations ... (so) I began talking about the medical effects of radiation upon testicles. Suddenly, you could have heard a pin drop ... The meeting ... (sent a) telegram to the prime minister ... (voicing) deep concern about uranium mining."

Encounters such as this escalated in number and prominence. the crusade invaded Helen's heart; she wept when she spoke of "children born and unborn who would sicken and die from our uranium." She wept over "my own powerlessness and frustration at knowing the truth but being overruled by ignorant, powerful men – who had never borne a child or cared for one as it died." Bitterly, she berated those

> "who make decisions ... but understand nothing of medicine or biology ... and businessmen and politicians who ... are scientifically illiterate ... (but) because I'm a woman, I'm easy to dismiss as 'emotional'."

Attending a convention of the Australian Pediatric Association in the country's capitol, Canberra, she persuaded nearly every physician to sign a newspaper advertisement describing all the dangers from uranium. The public was deeply impressed.

The Caldicott family returned to Massachusetts for a sabbatical year in 1975. Helen was far more poised and self-confident than on her previous employment at Harvard. "I was irritated to find so many ambitious and egocentric young men ... (who) would push me out of the way to peer down the gastroscope or up the colonoscope ... after four months, I got fed up and started behaving the Harvard way; only then did they seem to treat me with respect."

Having moved easily among Australia's senior union officials, she thought to do the same in the U.S., procuring an appointment with George Meany, president of the multi-million-member AFL/CIO. The mogul couldn't see her, but authorized another senior officer to receive her. He took her to lunch. She enumerated her arguments, her facts and figures. Seemingly impressed, he said he would try to arrange for her to speak with the organization's Energy Policy Committee; she would hear from him soon. She did, several weeks later, when he was in Boston. The invitation to appear before a union committee was to have a prelude, she learned, in his hotel bedroom. She fled.

At Ralph Nader's "Critical Mass" conference in Washington in 1976, she so impressed delegates from California that they invited her to campaign for Proposition 15 on the state ballot banning use of nuclear power. She spoke to groups all over the state for two weeks. The referendum was narrowly defeated by the overwhelming weight of industry advertising.

Her career as an anti-nuclear activist apparently did not affect Helen's reputation as an expert in her medical specialty. She was invited to join the Harvard faculty, fulltime, in the hospital's cystic fibrosis clinic. Husband Bill simultaneously was to become director of pediatric research in radiology. The offers were too good to be refused. They had six months in which to transfer. In Australia, they sold their home; packed their possessions and furniture, "made the rounds" of family and friends to say goodbye. As though she hadn't been away for a year, Helen picked up agitating within the union movement to nudge it toward dealing with the uranium-mining question honestly.

Nuclear policy had moved steadily upward on the public agenda. Export of uranium "yellowcake" triggered wharf strikes in Sydney, Melbourne and Darwin. Though uranium-mining was critical to the Australian economy – thousands and thousands of families derived their livelihoods therefrom – Helen's native state, Victoria, enacted legislation forbidding exploration, mining, and processing of uranium. Disposal of nuclear waste was become worrisome. The federal government, however, was still dodging the issue, promising that Australian uranium would not be employed in the production of nuclear weapons, though how this sanctimony would be policed was not explained.

Returned to America, Helen was invited by a publisher to write a book. "Nuclear Madness" became a best-seller; she was vaulted into national prominence. Invitations to speak multiplied. As did appeals to participate in demonstrations, which she did with increasing frequency.

Helen Caldicott's passion attracted like-minded personalities to her side. In Australia, she had birthed the Movement Against Uranium Mining. In her Newton, Massachusetts, home, she organized what became Physicians for Social Responsibility. Busy as doctors are, keeping up with their practice, keeping up with medical literature, keeping up with hospital service, they, in the main, largely held aloof from the anti-nuclear movement. Yet, Helen and her cronies reasoned, who should know about the "side effects" of uranium mining, plutonium enrichment, nuclear experimentation, nuclear dust fallout, and generation of nuclear-power? These consequences represented a greater threat to public health than any plague in history! The group produced an advertisement for the New England Journal of Medicine, enumerating the medical dangers of nuclear power. The responsibility of health-care professionals to become involved was underscored by the endorsement of nationally-known physicians who had been canvassed.

Like so many other worthy causes, the appeal to common sense would quickly fade from memory – except that, by incredible coincidence, it appeared in the "Journal" March 29, 1979 – the day after the Three Mile Island (nuclear plant) meltdown! Overnight, the nation – the world – shook off its psychic numbness. No longer was danger from nuclear power "theoretical." No longer could statesmen continue to ignore the subject – even if they themselves were naïve. (To calm the nation's hysteria, the President of the United States walked through the rubble of the TMI plant – and into the reactors themselves – without lead shields.)

Radiation monitors had been knocked out at the beginning of the accident, so there was no record – nor could there ever be any estimate – of the amount of radiation released by the disaster. Nor any prediction of what consequences may emerge during the millennium of half-life in nuclear discharge. In her speeches, Helen flogged these facts heavily. But in so doing, she provoked grumbles from some in the inner circle of the Physicians for Social Responsibility, who felt that she was diverting attention from their principal concern. She was told curtly "You shouldn't be talking about nuclear power; nuclear war is the really important issue to tackle."

Unlike Helen, her colleagues rarely ventured out of their offices and classrooms around Cambridge, Massachusetts. Helen had a clearer idea of what Americans in the heartland were worried about. She continued her jeremiads. She spoke from the steps of state capitols. She debated on public television. She spoke in Cuba. She spoke at Hiroshima.

In Moscow, she met with "a delegation of doctors, whom I was eager to persuade to start a Soviet-style PSR." She was the only doctor in the U.S. group at that meeting, hence presented her risk-to-health arguments with professional authority. She was later criticized by her colleagues for "dominating" the meeting. She brushed aside all criticism; she was launched into becoming an international figure. What she spoke about is what people came to hear. She spoke in Amsterdam, Brussels, Hamburg, Berlin, Ireland and all over Britain, launching there a counterpart to PSR, bannered "Medical Campaign Against Nuclear Weapons." She criss-crossed the country , helping to organize new affiliates for the British organization. But the same conflict arose. The eminent immunologist who was elected Britain's national president opposed dealing with nuclear power rather than exclusively opposing nuclear weaponry.

The medical profession in England and Scotland – as in America – was far from convinced that radiation fallout was a serious disease-threat to public health. Many researchers were hard at work, of course; their efforts would eventually bear bitter fruit – years later. But at that time, proofs were not conclusive. There is, everywhere, a heavily-financed nuclear industry with enormous investment in reactors. Pure science moves slowly. In Britain, a "Medical Research Council" had, since 1955, been investigating the possible connection between radiation and bone-marrow cancer. "Possible" is the key word. Until its Report was conclusive and could be deemed unassailable, it was just one more alarmist theory. Meanwhile, "experts" on the other side were available for hire to trumpet the standard rhetoric that these ideas were "unproved."

While Helen was abroad, back home the resistance movement grew in the PSR, at its roots a combination of male chauvinism and envy of her worldwide prominence. It was strongly argued that the movement, if it were to acquire a permanent political clout for peace, had to represent more than Helen Caldicott's convictions, however passionate and successful her missionizing.

In December of 1980, a rump group within the PSR organized International Physicians for the Prevention of Nuclear War, joining with three doctors from the Soviet Union. It launched a membership drive. Helen resented the competition for members and for funding, but refrained from direct conflict. She was swimming at the crest of a tide, having resigned from Harvard to devote all her time to the crusade. During these intoxicating years, she stormed around the country, appearing on television, making speeches before all sorts of groups, participating in protest meetings, recognized as the pre-eminent spokesperson for the anti-nuclear movement. A film was produced and entered in the documentary category of the Academy Awards; it won. Meryl Streep sponsored its exhibition on Broadway, with Helen's name on the marquee. The Justice Department banned its distribution. (This infringement upon the constitutional guarantee of freedom of speech was contested up to the Supreme Court, where the verdict was upheld.)

Helen was "profiled" and interviewed in national magazines and newspapers, and on national television. She had especially great influence on Hollywood women – most of them mothers – like Sally Field, Candice Bergen, Goldie Hawn, Lily Tomlin, Debra Winger, Jane Alexander, and Margot Kidder. Thus was born "Women's Action for Nuclear Disarmament." Through female show-business luminaries, she became friendly with and enlisted the support of male stars, too: Jack Lemmon, Kris Kristofferson, Paul Newman, Robert Redford. Even former "hawks" like William Colby, retired head of the C.I.A., were converted. Through Ms. Kidder, a Canadian, she met and educated Pierre Trudeau, Prime Minister of Canada.

Trudeau booked her to speak before the Canadian cabinet – face-to-face with James Schlesinger, U.S. Secretary of Defense, who had come to Canada to justify U.S. cold-war foreign policy. His catechism was simple – "if people had to be killed in the name of anti-communism, so be it ... Canada (has) no right to interfere with or even comment on American foreign policy." When it was Helen's turn, she bluntly told the men who governed Canada what they could expect to deal with, after nuclear attack:

> "Enormous pressures will create winds of up to 500 miles an hour, causing hundreds of thousands of injuries ... (it) will literally pick people up off the pavement and suck them out of the reinforced concrete buildings ... converting them into missiles traveling at one hundred and twenty miles per hour. When they hit the nearest wall ... they will be killed instantly from fractured skulls, brain trauma ... and injuries to internal organs ... Windows will ... shatter into millions of shards of flying glass ... (that) will penetrate human flesh ..."

She then turned to Schlesinger: "Every human life is as precious as yours, and you don't kill people in the name of foreign policy."

When the conference ended, Trudeau invited six countries to a conference in Canada at which the first international alliance against nuclear war was born. The "Five-Continent, Six-Nation Peace Initiative" called for an immediate and unconditional nuclear freeze. Though the U.S. Administration had voiced support of negotiations toward a nuclear-détente, Helen quoted (former President) Richard Nixon: "Don't listen to what we say, watch what we do."

Senators Ted Kennedy (Democrat) and Mark Hatfield (Republican) introduced a "sense of the Congress" resolution calling for nuclear freeze. Polltakers reported approval by more than 60% of the American electorate. Congressman Les Aspin, (a leading authority on military matters, later Secretary of Defense, further proposed a resolution to bar funding for development, testing, acquiring or employing nuclear weaponry that would contravene "freeze" regulations in the SALT treaty, then being negotiated between the U.S. and the Soviet Union. By the time President Reagan countered with his "Star Wars" program, the approval-rating of 60% for a nuclear-freeze had become — in less than a year — 80%. National media found irresistible the conflict between a democratically-expressed national will in conflict with Administration cold-warriors. A television network produced "The Day After," depicting medical and sociological aftermath to the dropping of a nuclear bomb on an American city. Helen Caldicott persuaded the Speaker of the House of Representatives to re-broadcast it into every Congressman's office.

During her brief stay in Washington, Helen met President Ronald Reagan's daughter. The young woman arranged a private meeting with her father. Flat out, Reagan said "I believe in building more bombs." He was convinced that the Soviet Union was "stronger than America, (and) wanted communism to take over the world." He pulled from his pocket "notes … (that) the freeze campaign was being orchestrated by Russia … we were KGB dupes." Helen rebutted: the data he identified as "from my intelligence files," she informed Reagan was, in fact, taken directly from an article in "Reader's Digest."

On a nationwide speaking tour, Helen interviewed General Bennie L. David, commander-in-chief of the Strategic Air Command at Omaha, Nebraska. From him, she learned that — behind the scenes — Administration policy had shifted from accepting the de facto SALT standoff between equals who were equally constrained by the reality of mutually-assured destruction, and was working on a first-strike theory — " … winnable nuclear war."

By then, nearly every medical body in America endorsed resolutions against nuclear war. The "New York Times" hailed PSR as having done "more than any other group to thrust the nuclear issue under the public eye." Thus lauded, there was no holding Helen; her militancy ratcheted upward. She determined to confront Goliath — the "military-industrial complex." This was the phrase used by Dwight Eisenhower in his farewell address after two terms as president, warning the American people that it represented a threat to democracy. Helen now revived that warning. She went on the attack, charging that it

> "was no longer enough to move to tears by describing the symptoms, the medical consequences, of nuclear war … I therefore encouraged audiences to look at the relevant institutions which had created the nuclear monster …"

"The PSR executive worried (that) … we would lose the support of our conservative colleagues …" In effect, she thereby forced them into action. An effort was made to demote Helen to a subsidiary position in the "table of organization." Meetings of the directorate grew heated. The slim woman, no longer young, was already exhausted from travel and public appearances. Having debated Richard Perle, an especially eloquent Administration "hawk" at the American Newspaper Editors Association, she was next scheduled to debate Edward Teller, "father of the H-bomb," before the National Press Club. Her colleagues demanded that the PSR statement of their position be written by them. Helen warned: "You clip my wings and I'll flee."

The wing-clipping went ahead notwithstanding. A management-consultant report was introduced that ostensibly was to map the PSR future. The introductory paragraphs about Helen's role were fulsome, hailing her "vision, her creativity, her commitment, her ability to inspire and activate, her brilliant intuitive sense of strategy and her drive to achieve what at first seems impossible … a key factor in the organization's success and a major reason why the organization has come so far so quickly." Then came the inevitable "but," the alarm-bell that in "any organization, a point is reached where a charismatic leader can become overpowering rather than empowering."

September 19, 1983, Helen chaired her last meeting of the PSR Board as president. 16 of the 18 members voted her out of office. A few months later, she resigned from the organization she had mothered. A revised history of the PSR was issued, in which Helen's name was downplayed. "I was devastated ... the effective tool I had helped create to move the world toward nuclear disarmament had been kidnapped from under my wing." She notes with sad resignation that, in "Man and His Symbols," Carl Jung concludes "leaders are always betrayed or destroyed."

A chapter ended. The story didn't. Helen went to work in behalf of the Women's Action for Nuclear Disarmament; it soon grew to equal size with PSR. A national program was launched to get women knowledgeable about — and involved in — political affairs (only 2% of U.S. Congressional seats were then held by females). Helen wrote another book, "Missile Envy," in which competition between the superpowers was likened to adolescent posturing, acting out the male obsession with "winning." At a national convention of the Women's Conference on Preventing Nuclear War, at the Capital Building in Washington, she described the defense budget as being a $300 billion pork barrel at which legislators greedily fed, and gestured around the noble building: "This place is full of corporate prostitutes!"

Meltdown of the nuclear reactor at Chernobyl in Russia re-aroused her rhetoric:

> "800,000 children ... (are) at risk for developing leukemia ... many babies in the fallout area ... (have been) born without arms or legs and with other gross deformities ... 160,000 children below the age of seven ... (are) at risk for developing thyroid cancer from radioactive iodine ... 2,697 villages with a combined population of 2 million ... (are) seriously contaminated ... food grown in contaminated areas will be radioactive for thousands of years ... a total of 3.5 million people (are) at risk for cancer or leukemia."

Accused of being an alarmist, a scaremonger, "I lacked the strength to effectively fight back." She was bone-weary from flying around the world, responding to appeals from this group or that; tired of trying to be a mother of three, a wife, a home-maker — simultaneous with her career as a public figure. She became aware of a need for independence: "I had repressed many of my own feelings in order to live compatibly with Bill ... there was ... (growing) distance between us, I realized that I was very unhappy with him." Apparently, he was equally unhappy with her. August 6, 1987, just before Helen's 50th birthday, Bill told Helen that he had been involved for years with another woman, a friend of Helen's, in fact. The Caldicotts were divorced.

Helen withdrew from the international arena, returned to Australia, "alone, worthless, unloved." A national magazine interviewed her, the cover headlining "My Husband Left Me for Another Woman." As always, she was outspoken. Her two sons and daughter, now grown, were supportive. But they were launched on their own lives, rearing their own families. She became an environmentalist; continued to campaign for underdog issues; participated in political life. She ran for public office — unsuccessfully, having admitted in the course of campaigning that she believed in population control.

Can we assess the significance of Helen's career? Obviously, the universal revulsion toward nuclear experimentation and saber-rattling should not be credited to her sole efforts. Equally obvious, her leadership in motivating millions to feel that way cannot be overlooked. What can a single individual do about influencing the course of history? Eleanor Roosevelt, wife of President FDR, became famous for urging idealists not to curse the darkness, but to light a candle. Even one.

Helen Caldicott now lives in the United States and has returned to the practice of medicine. She signs off her autobiography, "A Desperate Passion" with reassurance that "my spiritual journey continues, as does my sense of joy. Life has never been sweeter or more precious to me, and I look forward to its next state."

She invites use of her website; her internet address is: www.mindspring.com/-hcaldic.

BENJAMIN CARSON, M.D.
From ghetto to
Worldwide fame

Early morning Saturday, September 5, 1987, at the Johns Hopkins University Hospital in Baltimore, an operation was scheduled to separate "Siamese-twin" baby boys, joined at the backs of their heads.

It was not the first time that surgeons had worked on twins joined at the head. Always before, one child had been sacrificed in the effort to repair and save the other. This time, an attempt would be made to save both – as the parents pleaded. Neurosurgeons elsewhere in the world who had experience in separating "Siamese-twins" declined to try because the infants shared a major vein in the circulatory system. A comparatively young doctor at Johns Hopkins thought otherwise.

Ben Carson devised a plan that was unique in pediatric neurosurgery. He intended to use a heart-lung machine to divert blood from the brains during the critical period of work within them.

> "I had come to the conclusion, after doing a lot of research, that the reason … Siamese twins didn't survive is exsanguination (blood draining). Hence, if we cool the body until the heart stops, pump all the blood out … (we'll be) able to operate … combine some of the neurovascular techniques that we have for separating and reconstructing vessels (during) hypothermic arrest … It had not been done, previous to that time … (It was) a matter of taking information that had already been derived and applying it to a problem to which it had never been applied before …"

The problem was simple; the solution, complex. The infants had but one complete vascular system connected to the two brains; and only one "superior saggital sinus vein." It was intended to fashion a second complete vascular network. New vessels would be created from pericardium (heart-covering) material, which would be excised in a preliminary procedure.

Except for their cranial anomaly, the two boys were normal in every way, and functioned separately as individuals. As the backs of their heads were fused, they faced in opposite directions. Unless the operation was successful, they would be bedridden for the rest of their lives.

At two adjoining tables that September morning stood Donlin Long, Chairman of Johns Hopkins' Department of Neurosurgery, and Benjamin

19

Carson, Chief Pediatric Neurosurgeon. They were surrounded by a small army of assistants, 70 in number: pediatric anesthesiologists – seven of them; neurosurgeons – five; cardiac surgeons – two; swarms of nurses and technicians.

The heart-lung machine was normally used for cardiac surgery, not neurosurgery. By diverting blood circulation in the tiny bodies during the period Carson and his colleague would need for reconstruction work in the two brains, extra time could be gained.

To guard against bypass-machine malfunction during this measure, resulting in hypothermic arrest, the operating room had been equipped with an extra set of bypass machines. Extra sets, too, of all other equipment that might be needed. The O.R. had also been wired with backup electrical circuitry to provide power in the event of interruption or overload. Carson's team had practiced every step in the surgery, using dolls with velcro – joined at the head. Notations and observations, even the smallest incidents, had filled a large book. Every team member had committed to memory where they were to stand and what their responsibilities embraced in every step of the unprecedented operation.

It commenced at 7:15 a.m, when cardiac surgeons anesthetized the babies. Catheters were inserted in all major veins and arteries, to act as sensors transmitting vital – sign information throughout the operation. Carson and Long then cut into and peeled back the babies' scalps, and drilled into their craniums. Pieces of bone from the fused area were cut out and preserved, to be used later for reconstruction of the skulls – if the operation were successful., They next carefully peeled back the "duras," a contorted and corrugated covering that conforms to brain tissue. An abnormal artery that connected the two brains was sectioned. Adhesions joining the two brains also had to be sectioned.

Veneous sinuses, branches of the vascular systems, normally merge at the "torqula," a junction-cluster-like branches joining a tree trunk. In a baby's brain, this knotty mass is usually about one inch in diameter. It was necessary to separate and realign the filaments, adding new vessels as needed to create fully independent vascular systems.

But instead of occupying an area of approximately one inch, the twins' torqula was attenuated almost down to juncture with the spinal cords. A great flood of veneous blood gushed forth.

In the painstaking rehearsals that the team had performed, five minutes had been considered more than ample to accomplish the torqula unraveling – while the boys' blood flow was diverted from their hearts through each of the machines. During "hypothermic arrest," body temperature can be cooled to 68 degrees to slow circulation significantly and control bleeding. However, that procedure, with babies this small, could be relied upon to keep life processes functioning for a maximum of one hour, before damage to brains and tissues would begin. Under intense time-pressure, Carson and Long each worked within a tiny skull, separating the infinitely small vascular passages. In addition, Carson had to construct a critical addition – the new saggital vein. The clock ticked remorselessly. If they could not complete the intricate assignment within the 60-minute deadline, they had ordered disconnect of the heart-lung machine and resumption of conventional blood flow. In the resulting hemorrhage, both babies would likely die.

During this breathless period, the cardiovascular specialists at the surgeons' sides were supplying new blood vessels, fashioned from the set-aside sections of pericardium, each sized and shaped to fit the space available. When they had all been sewed into place, barely one minute of the hour remained. The heart-lung bypass machines were disconnected; the boys' hearts resumed pumping through the now-complete, separated, vascular systems.

The next emergency struck: massive bleeding, a cataract. The doctors and nurses sopped and sponged. As fast as the life-blood of the babies was spilling out, units of whole blood were being pumped in. Minute after minute, the bleeding was seemingly unstoppable. The hospital's blood bank was exhausted. Emergency calls went out, and a fresh supply was rushed from elsewhere in Baltimore to Johns Hopkins. After nearly three hours, the flow slowed. During the three hours, 60 units of blood had been transfused – more than replacing the normal total blood volume in each of the two bodies.

The plastic surgeons had meanwhile been piecing together patches of scalps and sewing them into place. This, too, was under time-pressure, to confine the swelling brains within the craniums.

Carson and Long had been on their feet from early morning Saturday to early morning Sunday: 22 hours. As had the throng around them. All had rotated a few minutes at a time for food and comfort calls. At a hushed press conference held minutes later, Carson gave full credit to "incredibly competent people" for having made successful "what was perhaps the most complex surgical feat in the history of mankind."

A postscript is warranted. Donlin Long was a highly-experienced veteran, Chairman of the Neurosurgery Department for 14 years. Standing at his side, his junior, Benjamin Solomon Carson, was a slim, youthful figure, who had earned his M.D. a scant seven years before. He is a Black, one of the few of his race in pediatric neurosurgery, and at 34, was one of the youngest in the world.

Not an easy achievement for a ghetto kid. His mother had been married at 13, one of 23 siblings in a desperately poor family. His father was a bigamist, dividing his time with a second wife and family. He disappeared from Ben's family circle when his wife divorced him – abandoning responsibility to support her and his two sons when Ben was eight.

Which started them on a steep spiral into deep poverty. They had to give up their home and move from Detroit to Boston, where they were taken in by his mother's sister. Sonia, the boys' mother, worked three menial part-time jobs to keep the family unit fed and clothed.

They lived for two years crowded into tenement rooms. The building was surrounded by a wasteland of debris, in which rats ("the size of small dogs") bred – and occasionally invaded the building. Snakes, too. Roaches infested every room impervious to insect-control measures. Ben recalls:

> "winos and drunks flopped around the area, and we became used to seeing broken glass, trashed lots, dilapidated buildings, and squad cars racing up the street...
>
> "I was an adult before I discovered where Mother went when she "visited relatives." When the load became too heavy, she checked herself into a mental institution ... plunged ... into a terrible period of confusion and depression ... Usually she was gone for several weeks at a time."

The boys' education had not only been interrupted by the move east, but the level of instruction in their new school was so low that "passing" grades were really fakes. Ben's classwork and performance in quizzes and exams were abysmal. "I had no competition for the last place in my class." He – and everyone else – thought he was a dunce.

> "I didn't understand anything that was going on ... I think my poor record reinforced my general impression that Black kids just weren't as smart as white ones."

Not till he was 11 and in the fifth grade was it discovered that he had poor vision. The doctor told him, "Son, your vision is so bad you almost qualify to be labeled handicapped." He was fitted with eyeglasses by the school system – and began the long climb toward literacy.

Mrs. Carson now took an active role in the boys' education. She told Ben that for math,

> "you're going to ... memorize your 'times tables.' I only went through third grade, and I know them all the way through my twelves ... you're not to go outside and play after school ... until you've learned those tables."

"Within days after learning my times tables, math became so much easier ... Nobody laughed or called me the dummy in math anymore ..." So the boys' mother started the next phase.

> "I've decided you boys are watching too much television," she said one evening, snapping off the set ... "You boys are

going to the library and check out books. You're going to read at least two books every week. At the end of each week, you'll give me a report on what you've read." (Little did the lads know that she barely could read what they wrote! Years later, she earned a high-school equivalency.)

The boys were rebellious. "All my friends were outside playing basketball, baseball, football, dodge ball, just having the best old time, and there I was, in there reading. But by recycling spare time, slowly the realization came that I was getting better in all my school subjects ... By reading so much, my vocabulary automatically improved along with my comprehension ... I had to take those words and make them into sentences, so I learned grammar and syntax. All of the sudden, I was getting good marks on my English papers ... within the space of a year and a half, I went from no competition for last spot to no competition for first spot in my class."

The student body in his school at that time was mostly white; Carson became aware "that my intellectual growth helped to erase the stereotypical idea of Blacks being intellectually inferior." But it was at this age, too, he was accosted by whites who threatened bodily harm for walking in "their territory." Though the Carson brothers were the fastest athletes on the school football squad, they gave up the sport when white adults vented anger at their superior performance on the field ... "We never went back to practice."

(Racial prejudice does not end when one is young.) After Carson was already on staff at the Johns Hopkins Hospital, a white woman patient told his Chief, "I do not want a Black physician in on my case." Whereupon, Carson reports with a chuckle, "Dr. Long had a standard answer given in a calm, but firm voice. 'There's the door. You're welcome to walk through it. But if you stay here, Dr. Carson will handle your case'." Ben admits that a number of times he has been mistaken for an orderly, a char, or messenger. "It's all right because I realize most people do things based on their past experiences ... I could hear Mother's voice in the back of my head, saying ... 'Some people are ignorant and you have to educate them'."

There was a more serious problem plaguing the boy's adolescent years – false pride. He became conscious of fashionable clothes, and ashamed of the worn, un-stylish garb his mother brought home from clearance sales. So she invaded the austere family budget to provide high-style shirts, trousers and shoes. (This phase lasted, Dr. Carson laments, for years – until he joined the R.O.T.C. in his tenth grade, and was required to wear a uniform three days a week.)

During these sensitive years for a teen-ager, he was also ashamed of his family's poverty.

> "I felt the stigma of being poor most acutely because I didn't have a father ... we received food stamps and couldn't have made it without them. Occasionally, my mother sent me to the store to buy bread or milk with the stamps. I hated to go ... every time I left the house with the stamps burning in my pocket, I worried that someone might see me ..."

This fear of being derided lasted for years, diminishing gradually as affirmative emotions took over.

The boy loved music, and joined the band. The school's music teacher, however, "encouraged my academic pursuits. He saw that I had musical talent, but he told me, 'Carson, you have to put academics first. Always put first things first'."

The amazing mother was also determined that the boys should be exposed to culture. She memorized poems and aphorisms, and drilled the boys in them. (Carson is still word-perfect in reciting Robert Frost's "The Road Not Taken." Asked in an interview what his mother's influence has had on him, Dr. Carson replied: "There's an excellent chance that (otherwise) I wouldn't even be alive, considering the large number of individuals in my high school graduating class who are dead already.")

> "I started going downtown after school to the Detroit Institute of Arts. I walked through the exhibit rooms until I knew all the paintings in the main galleries ... Next, I had to learn about classical music ... I'd be out on the lawn digging up weeds ... my portable radio playing classical music. That was considered strange behavior for a Black kid in Motown ... I read books about operas and understood the stories."

(Fortune works in mysterious ways. Years later, he applied at an interview for admission to the neurosurgery program at Johns Hopkins Medical School. Only one or two out of a multitude of applications would be accepted. In the course of the interview with the head of the program, a European, it came out that Ben had been in the audience the night before at a concert which, coincidentally, was also attended by the older man., Invited to comment on the performance, Ben made pertinent observations. He was admitted to the program. Maybe he would have been, anyway – had he not been at the concert and discussed the performance knowledgeably. But it didn't hurt ...)

During Ben's high school career, he volunteered to work as a lab technician for the biology teacher. He was loaded with scholastic honors and finished third in his class. In those days, as today, students applied for admission to multiple colleges. The problem was, that each application had to be accompanied by a $10 fee. Ben had only $10 to spare, so he applied to but one, Yale University.

> "They offered me a 90 percent academic scholarship. After less than a week on campus, I discovered I wasn't that bright. All the students were bright; many of them extremely gifted and perceptive. Yale was a great leveler for me, because I now studied, worked, and lived with dozens of high-achieving students, and I didn't stand out among them ... The professors expected us to have done our homework before we came to class, then used that information as the basis for the day's lectures. This was a foreign concept to me. I'd slid through semester after semester in high school, studying only what I wanted, and then, being a good crammer, spent the last few days before exams memorizing like mad ... it was a shock to realize it wouldn't work (at Yale) ... I did backslide a little, but never to the point of not being prepared ... I started learning how to study, no longer concentrating on surface material and just what the professors were likely to ask on finals. I aimed to grasp everything in detail ... I applied the same principle to all my classes ... I ended up with a fairly respectable grade point average, although far from the top of the class. But, I knew I had done my best and put forth the maximum effort."

Along the line, Carson determined to become a physician, and upon graduation from Yale in 1973 with a B.A. in Psychology, was admitted to the University of Michigan School of Medicine. "I loved being a medical student ... neurosurgery ... soon intrigued me so much it became a compulsion ... Everything in print on the subject became an article I had to read."

He completed the program at Michigan creditably. He charted his future around "my interest in neurosurgery (and) my growing interest in the study of the brain ..."

He was a resident physician at Johns Hopkins University Hospital from 1978 to 1982; during which time he doubled as senior resident for a year at the Baltimore City Hospital.

> "A highlight ... during my residency was the research I did during my fifth year ... in the areas of brain tumors and neuro-oncology (cancer) ... I learned how to take out brain tumors ... I learned to correct malformations of bone ... and to operate on the spine. I learned to hold an air-powered drill ... to cut

through bone only millimeters away from nerves and brain tissue. I learned when to be aggressive and when to hold back."

When his five years of internship and residencies concluded, Ben Carson – now married to Candy, also a med student – was invited for another year as senior registrar at a teaching hospital in Perth, in remote Western Australia. As one of the few qualified neurosurgeons for vast distances around, he performed a great variety of operations.

"I did a lot of tough cases, some absolutely spectacular ... For instance, the Fire Chief in Perth had an incredibly large tumor involving all the major blood vessels around the anterior part of the base of his brain. I had to operate on the man three times to get all the tumor out ... (he) had a rocky course, but eventually he did extremely well. I (acquired) more experience during my one year in Australia than many doctors get in a lifetime of medical practice."

Very soon after the Carsons returned to Baltimore in 1984 – with a baby son – Ben was named Chief of Pediatric Neurosurgery at Johns Hopkins Hospital, at 32.

"Many parents brought very sick children ... often traveling great distances. When I walked into the room (to meet them for the first time) more than once a parent looked and asked, 'When is Dr. Carson coming?' 'He's already here,' I'd answer and smile."

In 1985, Ben made history with an operation on a four-year-old girl, Maranda Francisco. When she was 18 months old, she had suffered her first seizure, diagnosed as an epileptic fit. They became increasingly frequent. By the age of four, the child sometimes suffered more than 100 seizures a day.

"so they started feeding her through a nasogastric tube ... Maranda was forgetting how to walk, talk, ... and she needed constant medication."

Carson met the family after the child had been seen by numerous specialists, over several years. "More than one physician mislabeled her a mentally retarded epileptic." He quotes Mrs. Francisco: "My daughter has been on 35 different drugs at one time or another ... Often they'd give her so much she wouldn't recognize me." The parents had come to Johns Hopkins despite being told that Maranda's condition was inoperable.

By then, it had been diagnosed at the Children's Epilepsy Center in Denver and confirmed at the UCLA Medical Center in California as not being epilepsy, but instead, "Rasmussen's encephalitis," inflammation of brain tissue. Fifty years before, a Canadian doctor by that name had diagnosed the disease as specifically afflicting only one side of the brain. Theodore Rasmussen performed dozens of operations over the next decades, completely removing the affected portion of the brain. Though others adopted the procedure, they gradually abandoned it in the face of discouragement over the low success rate and the many problems that followed surgery – in part because of bleeding and infections in the cavity left by removal of part of the brain.

Nobody was willing to undertake Maranda's surgery. They referred the problem to Dr. John Freeman, pediatric chief of neurology at Johns Hopkins. Freeman had previously been on the staff at Stanford University Hospital and there became familiar with a procedure known as hemispherectomy. This radical measure, removal of the affected half of the brain, is the Rasmussen technique. Freeman consulted Carson. Carson, who had never before done that procedure, assembled all the literature available on the disease and everything in print detailing experiences with the surgery.

> "I reasoned that she was having so many seizures that she
> had no life ... there was not really anything to risk."

If nothing was tried, little Maranda was sentenced to death – preceded by agonized suffering. Carson, at his first interview with the parents when they came to Baltimore, told them:

> "It's a dangerous operation. Maranda may well die ... (or
> after the operation) have significant limitations, including severe
> brain damage ..."

Informed of the risks, aware that the child's quality of life would further deteriorate toward an early death, the parents authorized the operation.

The surgery had been estimated as requiring five hours; it ran to double that time.

> "Maranda's brain was very inflamed, and no matter where an
> instrument touched, she started to bleed. It was not only a
> lengthy operation, but one of the most difficult I'd ever done."

Not to burden this account by again describing step-by-step the brain surgery, Carson's synopsis is offered:

> "Slowly, carefully, for eight tedious hours, I inched away the inflamed left hemisphere of Maranda's brain ... a millimeter at a time, coaxing tissue away from the vital blood vessels, trying not to touch or damage the other fragile parts of her brain ... Maranda lost nearly nine pints of blood during the surgery. We replaced almost twice her normal blood volume."

After the surgery in August, 1985, Maranda's recovery was slow. Motor coordination of fingers on her right hand was impaired. She walks with a limp – as she did before the operation. But there have been no more seizures. She took up tap dancing.

> "Children's brains have considerable overlap," Carson explains, so that "functions once governed by a set of cells in the brain are taken over by another set of cells. No one understands exactly how this works."

Hemispherectomy operations returned to the practice of neurosurgery worldwide after that. Carson logged 30 in the following years – the youngest patient barely three months old. Only one of the 30, operated upon at age five months, died. Whereas almost every one of the other children have seizures no longer; and have improved intellectually as a result of the surgery. Carson subsequently pioneered with prenatal brain surgery – operating inside the skull of an as-yet-unborn infant.

He works a 12-hour day, but reserves weekends for family life. "I grew up without a father and I don't want my sons (there are now three) to grow up without one ... My wife, my sons – they are the most important part of my life."

He expresses a "self-imposed obligation to act as a role model for Black youngsters ... the way to escape their often dismal situations is contained within themselves. They can't expect other people to do it for them. I believe that many of our pressing racial problems will be taken care of when we who are among the minorities will stand on our own feet and refuse to look to anybody else to save us from our situations ... You only have so much time in your life and you only have so much energy. So you have to select very carefully how you're going to spend that time and how you're going to spend that energy."

The quiet voice of

RACHEL CARSON
Launched a revolution

Technology produced so many wonders during the decade following the Second World War, a never-ending stream seemed to shower from heaven. Blood-plasma and electronics and antibiotics and labor-saving household appliances and all sorts of synthetic textiles and ...

The most important of them all, DDT. Dichloro-dipphenyl-ethane had been originally developed in Hitler's Germany as a basis for chemical weaponry. It had then been re-formulated as an insecticide by the Swiss Paul Muller, for which he was named a Nobel laureate. In the form of a spray, DDT was enormously valuable: it killed lice, controlled typhus and denied to malaria-carrier mosquitoes their marshland breeding grounds. It was quickly adapted to civilian uses in agricultural and forestry management applications. The chemical industry mushroomed, producing every year compounds that promised to kill more bugs more efficiently.

Then came a startling turnabout. Scientific journals reported outbreaks of diseases that were supposed to be derived from poisonous substances. The Food and Drug Administration and the National Cancer Institute sounded alarms. What was in process was "a triumphant vindication of Darwin's principle of the survival of the fittest ... (insects) evolved super-races immune to the particular insecticide used, hence a deadlier one has always to be developed — and then a deadlier one than that ... the chemical war is never won, and all life is caught in its crossfire." (Rachel Carson's words.)

Alarm bells were not welcomed by the chemical industry, grown to giant size. There were almost no controls on the compounding of poisons; or on combinations of them; or relative strengths; or dosage; or frequency of application. At the heart of many DDT formulae was arsenic, the oldest and most potent poison known to man. Products appeared on store shelves that were hundreds of times more toxic than DDT — fertilizers, fungicides, herbicides, insecticides. All were marketed in the name of "progress."

How to deal with a Frankenstein's monster? Status quo is usually recognizable as a retardant to scientific progress; but a hugely profitable status quo had become the protector of vested interests. The chemical industry was a favorite investment with hundreds of thousands of workers in chemical plants; combined, they represented a seemingly impregnable problem. No single individual, no scientific institution, no government agency dared to give battle. Except ...

Rachel Carson, at 50, in retirement from the Civil Service, took up the challenge. She was familiar with the evidence that chemical sprays were believd to cause evil consequences. She also was familiar with the rules of science that prohibited using "anecdotal" evidence as proof. Like most people, she'd heard that cancers and genetic damage in the families of Hiroshima and Nagasaki survivors had been discounted as "anecdotal."

The anti-DDT case was equally vulnerable as anecdotal. Clinical proofs were needed to connect effects with causes. Evidence so weighty it could not be ignored or deprecated. Rachel Carson knew it would be a difficult task. Her Ph.D. in biology didn't endow special credentials to conduct the research. She had no access to a laboratory, no institutional sponsorship, no "inside knowledge," no funding. No reason to stick her neck out, either, as she had no career to further, she didn't need money, she was comfortable in retirement.

Before her retirement from government service, she'd become a highly successful author of non-fiction books about marine life. "The Sea Around Us" was the first of several titles that earned international popularity. It had been pre-published as a condensation in "Readers Digest." She proposed to the editors of that magazine – with whom she was by now friendly — a thorough review of all that was then known about chemical poisons' effects upon every form of life. She would do all the "donkey work" herself and then sum up the findings in a long essay, to be published exclusively by the magazine. The publication, which had profited handsomely from Rachel's writings, turned her down. They were not in the business of upsetting apple carts.

Rachel was only briefly deterred. With no "advances" on royalties, no commitment from any publisher that her work would ever see print, no financial backing from any quarter, she worked on the DDT project for four years. She haunted dusty file rooms; reviewed back numbers of professional journals, regional studies by university and agricultural-extension services, exhumed laboratory reports and research abstracts, even delved into hospital records. She assembled and then selected, from the mountain of data, facts that had pertinence to each subject in her overall plan. To each, she added biological information to explain relevance. When appropriate, she peppered her narrative with historical footnotes. In this painstaking, tedious process, she wove warp and woof of an intricate fabric, fitting together seemingly disparate pieces like a gigantic jig-saw puzzle, until everything fell into place seamlessly. She thus organized an enormous body of knowledge, complete, coherent, comprehensive — and incontrovertible. DDT and its chemical cousins, thought for so long to be beneficial to human enterprises, were actually lethal — "not an occasional dose of poison which has accidentally got into some article of food, but a persistent and continuous poisoning of the whole human environment," she wrote.

In 1962, "Silent Spring" was published: a 250-page collection of essays. Each dealt with a different damage stemming from chemical substances. In each, she traced destructive interaction between chemical poisons and life forms. In one chapter, she explained the molecular structure of insecticides — hardly a compelling subject — until she related it to the specter of genetic fallout. "We are rightly appalled by the genetic effects of radiation; how, then, can we be indifferent to the same effect in chemicals that we disseminate widely in our environment?"

Discussing "Surface Waters and Underground Seas," she noted that, although insecticides are sprayed in very low concentrations, the poisons gain strength as they progress through the food chain and reach levels far beyond the limits of safety. Carcasses examined by the Food and Drug Administration (which she gave as an example) "contained concentrations of chemicals too dangerous for human consumption, even in minute quantities."

She salutes the lowly earthworm for its many services to nature — but laments its role as a carrier of poisons for birds. Spraying to save Dutch Elm trees from a blight spread by one type of insect did not save the trees, but killed beneficent insects that were staple nourishment for songbirds. Every year, she wrote, they wove fewer nests; every year, their nests contained fewer fertilized eggs. Eighty percent of the robin population vanished in one area. "Treatments" against Japanese beetles slaughtered small animals, too, including household pets.

In the "Rivers of Death" chapter, she described the massacre of millions of fish, which are especially vulnerable to chlorinated hydrocarbons, washed into streams, lakes, and even into oceans that are fed by contaminated bodies of water.

"Indiscriminately from the Skies" reminded readers that thousands of "surplus" airplanes had become idled after the war, as had thousands of experienced pilots. Many became auxiliaries to the chemical industry as free-lance "crop-dusters," who were paid by the gallons of fluids they spread, not the farm acreage covered. There were no controls on saturation levels. Farm animals died after foraging in areas that had been heavily sprayed. So did newborn calves dependent on their mother's milk. As for human consumption of the same milk, although the federal government mandated minimum pesticide residues in milk shipped by dairies – the rule covered only milk shipped interstate! Most farms then were small and sold their production to nearby dairies; there were almost no controls on milk shipped intra-state. (Carson didn't miss the opportunity to report that the FDA checked less than 1% of crop products that traveled in interstate commerce. She twanged every possible chord to muster a chorus for protest. Sportsmen were told that in sprayed areas, wild turkeys nearly disappeared, also bobwhite quails and other ground-nesters.)

In "Earth's Green Mantle," she reported the havoc wrought on Douglas fir forests in nearly a million acres of federal land sprayed with DDT by the U.S. government to kill budworms. As a result, the spider mite population went into an explosive growth stage — far more damaging to the trees than the budworms. She cited many other examples of nature's delicate balance being needlessly, heedlessly, upset by indiscriminate poison sprays. Some "target" insects were almost unaffected while their natural enemies, predators, were vulnerable. She described such an innocent victim: " ... almost invisible against a leaf ... the lacewing, with green gauge wings and golden eyes, shy and sensitive, descendant of an ancient race that lived in Parmian times." She reminds: "Most of us walk unseeing through the world, unaware alike of its beauties, its wonders, and the strange and sometimes terrible intensity of the lives that are being lived around us."

Such touches of lovely prose appear sparingly. Outrage and irony are more frequent. Spray-bombing fire-ants in the South, cropdusters killed off insects that had long controlled sugarcane borers — which are vastly more destructive to a cash crop then were the ants, at most a minor nuisance. The process was "...as crude a weapon as the cave man's club... (a) chemical barrage has been hurled against the fabric of life ..."

Every good narrative includes conflict. For which purpose, she provided, as an example, the U.S. Department of Agriculture. At all levels, government is a lucrative market. Thousands of chemical engineers work in government agencies and know their job security is linked to the war against insects. The pest-control section of the Agriculture Department in Washington was, at that time, eager to spread "eradication" campaigns in grazing areas. In the Washington building, research scientists were, however, publishing warnings that major ingredients in those very pesticides were hazardous to forage vegetation; severe collateral damage was done by the spraying. One of the states most concerned – Alabama – bitterly reported the program was "ill-advised, hastily conceived, poorly planned." Other states echoed the complaint.

Such episodes were commonplace. Carson explained: "There is no dearth of men who understand these things, "but these are not the men who order the wholesale drenching of the landscape with chemicals." Ninety-eight percent of the entomologists at work in the U.S., she pointed out, were employed to create new insecticides. The chemical industry provided grants to educational institutions, fellowships, and careers in the industry for the best. The two-way traffic between government service and private industry was brisk.

Near the end of the book, in a chapter headed "The Human Price," she quoted the Journal of the American Medical Association: "delayed effects of absorbing small amounts of the pesticides ... (can produce) biological effects ... (that are) cumulative." She added another warning, from the U.S. Public Health Service: "our fate could ... be sealed twenty or more years before the development of symptoms." She quoted Supreme Court Justice William O. Douglas: natural beauty is a patrimony, as valuable as "the veins of copper and gold in our hills and the forests in our mountains."

"Silent Sprint" first appeared as a three-installment abridgement in "New Yorker." For thoughtful readers, its effect was electrifying. No essay or story in the magazine's history had attracted so much mail before. Parts of it were read by legislators into the "Congressional Record." Justice Douglas termed it "the most important chronicle of this century for the human race." The revered scientist, Professor Loren Eiseley of the University of Pennsylvania, described it as a "relentless attack upon human carelessness, greed and irresponsibility."

Hundreds of thousands of copies were sold within the first few months.

Predictably, Rachel Carson became Public Enemy Number One to an oligarchy of agricultural, chemical and financial interests. She was vilified too by public officials who defended their special concerns and interests. She was guilty of being "antihumanitarian" and "hysterical," accused of representing "vociferous, misinformed ... nature-balancing, organic-gardening, bird-loving, unreasonable citizenry." She was derided as a food-faddist, a nature nut, a "fish-lover." America's largest producer of DDT deprecated Carson for writing not "as a scientist but rather as a fanatic ..." An executive from the Federal Pest Control Review Board sneered "I thought she was a spinster. What's she so worried about genetics for?"

Every form of society, ancient and modern, has had weaknesses. Capitalism, too. Spokespersons in important organizations too often "knee-jerk" react when their interests seem to be threatened. Sadly, even issues that involve public health and safety, if they are perceived to attack private interests, can evoke inappropriate responses. (Witness the behavior of America's tobacco company presidents — all of them — who were "sworn in" at a Congressional hearing, but then perjured themselves under oath — all of them.)

Precisely that amorality was evidenced when Rachel Carson exposed the deadly dangers inherent in the indiscriminate use of chemical compounds. An army of enemies attacked her veracity, motives, integrity. But her four years of hard work and her beautiful writing by then had mobilized a far greater army of friends. Public outcry over DDT poisoning arose all over the world. Educators, journalists, government leaders became environmentalists almost overnight. President John F. Kennedy appointed a task force to study her charges. She delivered data to his Science Advisory Committee. She also acquired skills in grassroots lobbying, with a heavy schedule of speaking engagements. She appeared as a witness at Congressional hearings and deeply impressed the legislators. Their report was issued May 15, 1963 and the very next day, a sub-section of the Senate Committee on Government Operations authorized review, overhaul, and strengthening of government restrictions pertinent to the manufacture and use of pesticides. With speed rare in democratic government — particularly rare when issues impinge upon the interests of big business — controls were created, groundwork laid for what became the Environmental

Protection Agency (EPA). It was a mighty victory. The war had been declared, fought and won - single-handed.

Nothing in Rachel Carson's previous career had promised such courage. She was not a zealot. She was not burning with the holy fire of discovery. Her life till "Silent Spring" had been conventional. She had been born in 1907 into a normal, functioning family, in a five-room frame dwelling on a 60-acre farm in Pennsylvania. She grew up happily in a rural environment; nature spread in all directions, offering infinite opportunities for adventure. There was "no time when I wasn't interested in the out-of-doors and the ... world of nature," Rachel reminisced years later.

Books were the sole luxury in the Carson household. The children were read to, nightly, and were expected to discuss with their parents what had been read — not only the narrative, but the moral message of the story. Otherwise, there were no distractions from real-world chores and natural-world delights.

Rachel early showed an interest in writing. She created her own stories and, at age ten, submitted one in a contest sponsored by a children's magazine. She won the prize, and ten dollars. Predictably, she chose English as her "major" when she entered a women's college in Pittsburgh on scholarship. It was for $100, far short of what was needed. The annual tuition, plus charge for books, lab fees, room and board totaled $1,000. School officials canvassed friends whose loans paid Rachel's way through all four years. The obligation was paid off years later when a buyer was found for some of the Carson family land.

In her second year of college, science surfaced as an interest stronger than literature. The Carson family still lived on the farm. Rachel still spent every weekend and holiday there, still loved the woods, the fields and streams, wildlife, hiking, exploring. She transferred from majoring in liberal arts to a science curriculum. The absence of basic science courses from rosters in her first two years were an awesome handicap. Rachel doubled up in her class load and — at 21 — was graduated magna cum laude in zoology.

Zoology? What kind of career objective was that for a woman? Which didn't deflect the confidence of her sponsors. Rewarding Rachel's outstanding performance, they procured for her a six-week summer scholarship in the Marine Biological Laboratory at Woods Hole, Massachusetts. This was her first sight of the sea — and it was love at that first sight. What few hours could be pilfered from the library and the lab were spent wandering along the coast. The more she observed, the keener grew her sensitivities.

It's said that opportunity comes to those with a prepared mind. In Rachel's case, there was a generous dollop of luck too. When she enrolled as a graduate student at Johns Hopkins University — on scholarship, of course — she took classes with Dr. Raymond Pearl. It had the effect on Rachel of an epiphany: a connection emerged between love of nature and awareness that the life, the health — the actual survival — of all species and all creatures on this planet, are linked together; all are dependent upon the behavior of man. Dr. Pearl was the first scientist to warn that cigarette smoking was a health risk.

Carson became Pearl's assistant. In time, she was awarded a part-time teaching position in the University of Maryland, while pursuing the graduate degree in zoology. She continued there for years. America was still deep in the slough of the Great Depression. Sixteen million able-bodied men had no jobs — 25% of the workforce. Her father died. Her older sister, Marion, also died — at 40, leaving two orphan daughters. Rachel assumed responsibility for rearing them, and caring for her own mother.

The mid-30's were scary times. Drought that lasted for years turned huge areas of western states into a desert, a "dust bowl." Livestock died. Farmers were evicted from land their families had tilled for generations. Automobile workers on a hunger march were shot dead at the Ford plant in Detroit. Veterans of World War I who converged on Washington to beg for help, were driven away by army cavalry.

There were no job openings for freshly-minted zoologists.

But early in President Franklin Roosevelt's New Deal" administration, a new government agency had been created, the Bureau of Fisheries and Wildlife. Elmer Higgins, head of the Division of Scientific Inquiry, was a bureaucrat with an imagination much wider than the official requirements of his position. He conceived the idea of producing a series of short, educational radio features about marine life; and wrote the first few episodes. There were no pollsters in those days, no "listenership ratings." But no mail arrived, either, so there was no way to assess the interest of an audience — or even if there were one. What feedback he got from "insiders" was negative: the stories lacked zest. At this point, Rachel Carson appeared in his office seeking work however lowly. He asked "Can you write?" She reassured him, got the job on a trial basis. Her writing was so much better than his, her stories quickly attracted a following. She was confirmed in her first full-time employment.

With an entry-level Civil Service appointment, Rachel invited her mother and two young nieces to make their home with her in Maryland. The Bureau of Fisheries salary was not high — government posts in those days were ill-paid but coveted for the security they provided. Jobholders were expected to wait for a superior to die, or retire, before promotion could be expected. When an opening finally was announced, Rachel along with many others took the Civil Service examination. She, the only woman, scored highest, became the first permanent-appointment female biologist in that department.

Rachel's stint as a radio writer eventually ended. Higgins by that time had become so fond of her seven-minute narratives that he conceived the idea of assembling them under an "umbrella" introduction and publishing them as a book. The preface that Carson wrote, however, was deemed too "literary" for an official government publication, and he instructed her to rewrite it. The quality of her original effort, however, stuck in his mind. He suggested she send it as an independent essay to the "Atlantic Monthly" magazine. The magazine instantly bought and published it with the headline "Undersea." Twenty years after her first success as a free-lance writer, Rachel earned her second reward.

Being published in the "Atlantic Monthly" warrants more than a bald statement of fact. In America, only a handful of general-readership periodicals are as highly regarded. To have one's first submission bought is amazing. It is doubly amazing when the author is not a famous name, and the subject matter is neither timely nor otherwise of special importance. Marine science is not on anybody's "top ten" list of interests.

"Undersea" was applauded by professionals in both scientific disciplines <u>and</u> in literature. Even more important, lay readers loved it, prompting an offer from the publishing house of Simon and Schuster for a full-length book. She wrote "Under the Sea Wind," which reached bookstalls just before December 7, 1941. Critics' reviews were excellent, but book-sales plummeted during the war years. The "failure" hurt. Security in her office-work appointment was never more comforting; it continued to be of prime importance to the Carson family for many years.

A steady stream of publications issued from her pen, hundreds of thousands of words. The literary quality of her work was so distinctive that her essays and articles were reprinted as editorial features by many newspapers and magazines, and narrated on radio. Her career as a writer flourished, but did little for her bank account. Government salaries then would not, today, attract teen-age applicants to work in fast-food restaurants. Rachel's mother was in her 70's; the nieces were in their teens; one income supported them all. In due course, Rachel was promoted to editor-in-chief of the Information Division, involving a great increase of administrative work that diminished time available for writing.

As a boss, she was quiet and gentle, said a senior assistant later, but demanded performance, and was unshakeable in her work-ethic principles ... "like Gibraltar" he added. "She had no patience with dishonesty or shirking." Another key aide rendered what he thought was the ultimate accolade: "she was a very able executive with almost a man's administrative qualities."

A major attraction in Civil Service employment was vacation leave of six weeks per year. Rachel invested much of her free time at Woods Hole. During summer interludes, and on weekends during the rest of the year, she worked on "The Sea Around Us." Published by Oxford University Press in 1951, it became a runaway best-seller and is now regarded as a classic, never out of print. The canny publisher had seen to it that individual chapters appeared first in "Yale Review," and "New Yorker;" a condensation appeared in "Reader's Digest." The book remained on the "New York Times" best-seller list for over a year and a half. Countless readers worldwide who shared her spiritual affinity to the mystery and majesty of great watery expanses, found in Rachel the poet who defined their feelings.

On the crest of that wave of popularity, "Under the Sea Wind" — which had "failed" on its original release in 1941 — was republished in 1952. It, too, zoomed into the "best-seller" category. The "Times" remarked that having two books by the same author that were best-sellers at the same time was a "publishing phenomenon rare as a total solar eclipse."

Royalties were rolling in. At 45, after nearly 20 years of low-paid government service, Rachel was able to retire from the Fish and Wildlife Service. She built a summer cottage on the rocky coast of Maine. In 1955, she wrote "The Edge of the Sea" there, to celebrate "an ancient world. Each time I enter it, I gain some new awareness of its beauty and its deeper meanings, sensing the intricate fabric of life by which one creature is linked with another, and each with its surroundings." The theme she first sounded in that book — interlinkage of every form of life with the environment common to them all — reverberated in all her work thereafter. Not since Charles Darwin had science spoken to such a wide audience with such authority and eloquence. It penetrated the consciousness of thoughtful people everywhere.

The editors of "Women's Home Companion" recognized her charisma and authority. Though she was not a mother, the magazine asked her to write "Help Your Child to Wonder." She recounted her experience, mentoring her 5-year-old grandnephew during a summer on the Atlantic shore. Alas, like many important literary works published in periodicals, the essay "disappeared" from print after publication. It has now been republished as a short book, and belongs in every home.

It is given to few scientists to write with technical competence while avoiding jargon or specialist vocabulary so as to convey their information interestingly to a general readership. Even more rare is the scientist who rises above that very respectable level into literary distinction. Such a one was Rachel Carson. She was honored with a gold medal from the New York Zoological Society, the Burroughs Medal and the Conservationist Award from the National Wildlife Federation. She won the National Book Award for non-fiction. She was honored with a Fellowship by the Royal Society of Literature in England.

Life continued in its tranquil course until Rachel passed age 50. Destiny then intervened. Rachel received a letter in January, 1958 from a reader, expressing "her ... bitter experience of a small world made lifeless ..." and brought her attention "sharply back to a problem with which I had long been concerned." Rachel's tranquil years ended and the genesis of "Silent Spring" began.

Mortally ill with breast cancer during her last speaking campaign before Congressional action, Rachel Carson died April 14, 1964. Posthumously, the highest honor that can be bestowed upon a civilian, the Congressional Medal of Freedom, was awarded in 1980, recognition that

"She "fed a spring of awareness across America and beyond. A biologist with gentle, clear voice, she welcomed her audiences to her love of the sea, while with an equally clear voice she warned Americans of the dangers human beings themselves pose for their own environment. Always concerned, always eloquent, she created a tide of environmental consciousness that has not ebbed."

Since Rachel Carson's death, the movement she birthed has become a credo. Ecology is now a mandate respected by governments everywhere. The very same company that accused Carson of being "fanatic" now buys large advertisements in favor of full label information about packaged food products. The European Union now controls the manufacture and marketing of potentially hazardous chemical products.

Environmentally speaking, this sort of awareness of responsibility was unknown B.C. — before Carson. She also initiated a history-making shift in civil jurisprudence, really a revolution. The "burden of proof" that a product is safe has now been shifted to the producers of insecticides – off the shoulders of those who formerly had to establish beyond shadow of doubt that damage resulted from application of chemicals.

Almost 500 new chemicals are developed for commercial use every year in the U.S. alone. Two hundred of them are formulated as sprays, dusts, fogs, vapors, aerosols. One and all claim value, but one and all pose risks. The planet cannot defend itself against assault. The constituency formed by Rachel Carson can, however. It represents a movement broader, more vigorous, more influential than any since Thomas Jefferson wrote the "Declaration of Independence."

MARIE SKLODOWSKA CURIE
Opened the door to Science for women.

Silently, save for the drum-roll, a wooden coffin was carried by tail-coated dignitaries into the Pantheon in Paris, April, 1995. It was entombed among 69 predecessors in that memorial dedicated to "great men of France." But the coffin carried the remains of a woman, not a man; Polish, not French.

Mme. Marie Curie was "the first lady in our history to be honored for her own merits," said President Francois Mitterand during the solemn ceremony.

It was the last of many "firsts" for Marie Curie. At 25, after only 18 months of study in a foreign language, the immigrant scored highest among the 30 students who took the final exam for a degree in physical sciences – and was the first woman ever to earn such a degree at the Sorbonne. She was the first to see what no human eye had seen before – the radioactive pattern of a hitherto-unknown element. She followed this historic discovery by identifying radium.

She initiated the medical specialty we know today as radiation therapy, in treatment of certain cancers. She was the first woman to be awarded a Nobel Prize. (She was not initially named by the Nobel Committee, but her husband, Pierre, refused to accept the Prize alone, in behalf of work they'd done together.) Exactly 15 years after the day she registered for classes in the Sorbonne, she stood before a packed auditorium, small and slim, all in black, to deliver the first lecture given there by a woman. Exactly five years later, she was awarded a second Nobel Prize – this one for chemistry; the first woman to be twice honored. She was the first woman to drive an ambulance during battlefield action and the first to take x-ray photographs prior to surgery – under fire.

At the Pantheon ceremony, the 1993 Nobel laureate for Physics, Pierre-Gilles de Gennes, asserts that she "changed the face of the world" by unlocking the secrets of nuclear energy; she had initiated the search for the component within atoms that caused radioactivity.

It had been a long journey for a frail immigrant girl. Her mother died when Manya Sklodowska was 12, having given birth to five children while helping to support the family by giving music lessons. To make ends meet, her father Wladyslaw, a poorly-paid teacher, took in boarders. Manya was the youngest, so she gave up her bed, slept on a sofa, and did the housework late at night, after her schoolwork. Notwithstanding, she was awarded a gold medal for excellence in her studies.

At 19, she became a governess in a wealthy home 60 miles outside Warsaw. The plan was to provide money so her older sister, Bronya, could study in Paris to become a physician, a six-year program.

Shortly before her 24th birthday, in 1891, Manya Sklodowska departed her home at 16 Freta Street in Warsaw and entrained for Paris to visit her now-married sister. The journey would take nearly four days, so she carried food, drink and books. Also, a mattress, blanket, sheets, towels and a chair. (There were no seats in a fourth-class rail car, usually used for cargo and animals.) The traveler had saved pennies for five years and helped Bronya through school. It was her turn to be helped; women in Poland were not accepted beyond secondary school.

Upon arrival, she lived for a short while with her sister, Bronya, and brother-in-law, Casimir, both physicians. Then she enrolled for the fall semester at the University of Paris – one of 12,000. On the registration form, she used the French spelling of her name, Marie. She had been out of classrooms nearly 10 years. There were vast empty spaces in her knowledge. Though she could read and write French – most educated Poles did – she had trouble following the rapid-fire speech of the professors. Her courses in natural science required knowledge of advanced mathematics, which she didn't have. The lecture-hall was huge and poorly lantern-lit; the blackboard barely visible. So Marie arrived early for every class, in order to have a seat in the front row.

> "All my mind was centered on my studies. All that I saw and learned ... delighted me ... a new world opened ... the world of science."

Studying at her sister's home was, however difficult, as her relatives entertained Polish expatriates almost every night. The young Ignatz Paderewski, later to become a world-famous concert pianist and president of his homeland, occasionally visited and pounded on the battered upright piano. When not nostalgically singing Polish songs, the group would talk and argue till late. Socializing was a pleasure, but not productive. Two hours daily commute to school by horse-bus also was time and money she could ill-afford. Marie had to move.

She found a rooftop garret room in the Latin Quarter near the college. She had no heat or running water. She bathed from a washbasin, carrying water up the stairs.

She studied by the light of a single kerosene lamp; used a tiny alcohol burner to heat water for tea and boil an egg. Coal for the stove was used only on freezing nights. She slept fully-dressed, piling her entire "wardrobe" atop the single blanket. To avoid using her coal, she often studied in a nearby church. When it closed at 10 p.m., she returned to her room and continued working till midnight or later.

She might have ended as a tubercular like her mother. She fainted in the street one day, weakened by malnutrition, fatigue and exposure. Her sister nursed her back to health. When Marie regained strength, she returned to the college and took extra classes in summer school.

She was awarded a scholarship for another year of study, also a small fee to analyze the effect of different chemicals on the magnetic properties of steel alloys. For which, she needed access to laboratory facilities and testing equipment. A childhood friend from Warsaw came to Paris on her honeymoon and introduced Marie to her new husband, a physicist. He arranged an interview with the Chief of Laboratories at the school of Physics and Chemistry in the nearby University of the City of Paris.

> "When I came in, Pierre was standing in the window recess
> ... I was struck by the expression ... (he) gave ... His rather
> slow, reflective words, his simplicity and his smile ..."

Thus, at 27, Marie met 35-year-old Pierre Curie, sixth generation in a family of French scientists. His father was a doctor and a naturalist. From their earliest years, he took his two sons on long walks in the countryside, explaining the wonders of botany and mineralogy. Early on, Pierre exhibited signs of extraordinary intelligence. His mother taught him to read. His father's medical books became his research library. Nature provided his laboratory.

Finally, when he was 14, "formal education" started: a tutor was hired for mathematics and Latin. By 16, he entered the Sorbonne. By 18, he earned the equivalent of a master's degree in physics, and was appointed an instructor in the Laboratory of Advanced Studies at the University of Paris. Independently, he began to analyze how crystals reacted to electrical charges of varying voltage under different levels of pressure.

His older brother, Jacques, with a doctorate in mineralogy, joined in the research. Together, over several years, they published a series of essays updating the state of the art in crystallography. Professional recognition earned for Pierre, at 24, a teaching appointment in the School of Industrial Physics and Chemistry. There he was content to stay for 10 years, during which time his publications were attracting attention, even outside France. Thus it was that, in 1893, Pierre received a visit by Professor Kowalski of the University of Fribourg – who had heard of him before he came to Paris on his honeymoon.

Pierre Curie already had read and thought much about the wonders of nature, a serious student of bird-lore, of botany and butterflies, especially fond of frogs. Marie Sklodowska was an avid learner; they put their bicycles on trains that reached far into the countryside for long woodland walks. Pierre wrote "It would be a fine thing ... to pass our lives near each other, hypnotized by our dreams ..." In her biography of Pierre, Marie described: " ... our friendship grew more and more precious. (We) realized that (we) could find no better life companion."

41

They were an ideal partnership. As goal-oriented as Marie was, Pierre wasn't. She badgered him to work on his Ph.D. dissertation about the structure and growth of crystals. No teaching appointment in science appropriate for a person of his talents could be sought without it. Once set in motion, he had no trouble earning the coveted degree. (His thesis paper on the effect of heat upon the magnetic strength of metals continues to be valuable in design of communications equipment right into the television age.)

The lovers were married July 26, 1895. They set up housekeeping with one table, two chairs, and a bed. When their first child, Irene, was born in 1897, a few more furnishings were added.

Pierre's doctorate brought him a small increase in salary and – of greater importance – permission for Marie to work in his laboratory at the college. Within months of her baby's birth, Marie published her first scientific paper on the magnetic properties of steel. She started work on her doctorate, undeterred by the fact that, in all of Europe, no woman had ever before earned one.

Then history intervened. They learned of radiation, a remarkable discovery by the German scientist, Wilhelm Roentgen. Modestly, he named it an "X-ray" which passed through opaque bodies and exposed, onto a photographic plate, shadowy traces of skeleton bones. Scientists all over started to experiment with the phenomenon. French physicist Henri Becquerel produced radiation, an amazingly powerful ray, invisible to the human eye, from decay of uranium salts. On the threshold of one of the great discoveries in the history of science – he turned back! The images projected by his rays were inferior in clarity to Roentgen's; and the electrical charges they produced were too weak to measure on his inadequate testing equipment. He published his initial finding, but did not continue work on radiation.

But Pierre and Marie had on hand exactly the testing instrument which Becquerel lacked – a gold-leaf electrometer that the Curie brothers had invented years before! It could measure electrical currents in the air so Marie could measure the strength of uranium rays by measuring the strength of the electrical current they caused. She learned that the strength of those rays depended only on the amount of uranium embedded in the ore. The strength of the rays was unaffected by whatever other substance was in the ore sample – wet or dry, powdered or in chunks, hot or cold. With that device, Marie was able also to identify which of the 83 known elements were contained therein:

> "I employed ... a plate condenser, one of the plates being covered with a uniform layer of uranium or of another finely-pulverized substance ... The current that traversed the condenser was measured in absolute value by means of an electrometer and a piezoelectric quartz."

The process was painstaking and tedious.   One by one, uranium ore specimens, wired to a power source, were put into a recycled jelly jar.   When charged, the circuit ionized the air in the jar, and registered, on the electrometer, the charge which increased with the amount of uranium in the ore-sample.

One day, she saw on her spectroscope screen the ray-pattern of another element that generated radiation, thorium. It was so faint as to be discernible only by a microscope – but it proved radiation was not unique to uranium. "Two uranium ores ... are much more active than uranium itself. This fact ... leads one to believe that these ores may contain an element much more active than uranium."

From the nearby Museum of Natural History, she borrowed a supply of ore-samples. Day after day, week after week, she tested them; some days, hundreds of them – minerals, rocks, sand. The process confirmed that only thorium and uranium oxide were known then to produce radioactivity. Yet, if two elements registered radioactivity on her spectroscope screen, didn't that indicate something unknown was within them – and common to the structure of both?   She expressed this theory in the very first paper published under her name: "Radiations Emitted by Compounds of Uranium and Thorium."

The date was April 12, 1898. She was still nursing Irene, her first child.

Marie received a 3,800-franc prize for her discovery – the first woman to be so honored in a field of science.

During the marathon-testing that finally led to Marie's seminal paper, she had learned something strange, and tucked it away in memory.   Crude pitchblende, a residue of the ore from which uranium was extracted for use in textile dyes, <u>actually yielded more radioactivity than did pure uranium.</u>

Early summer 1898, Pierre put aside his own work on crystals and joined Marie in a tedious history-making search. They took turns, alternating on the microscope and journal-entries. Pieces of pitchblende were tested, fragment by fragment, retaining for further testing only those that generated the strongest rays on the electrometer.   They identified its pattern on the spectroscope when electricity charged the specimen in the ionized-air chamber.

Came the day in July 1898 when spectrography reflected a pattern completely unlike any other known element. Marie named the new substance "Polonium."   Like prospecting for treasure in a wasteland, it was thrilling to discover the first clue. Were there others?

The painstaking search continued. After hundreds more tests, a second new spectrographic pattern flickered on the screen – even stronger than Polonium. The pair coined the word "Radium." It was December 26, 1898.

But they had only discovered "tracings" of radium. They were a long way from knowing its potentialities. To go further required isolating and working with the real thing, radioactive pure radium salts. Back to work. Knowing that the lode was buried in pitchblende, they bought a ton of the slag from a factory in

Austria. It was delivered to the "new" laboratory they had leased on the Rue Lhomond – a shed with a leaky roof, drafty windows, dirt floor, defective plumbing, almost no heat. Wilhelm Ostwalk, a German scientist, described it as "cross between a stable and a potato cellar."

When they started the hunting process, they did not know how long it would take – or if it would be successful. One ton of pitchblende would require fifty tons of water and five to six tons of chemicals. The largest receptacle they could handle would only contain fifty pounds of pitchblende sludge. It would require hundreds of experiments to consume the mountain – over four years of eye-glazing, back-breaking work. They knew, from Marie's first experiments, which elements were natural to pitchblende. By identifying them, one by one, the residue would have to be the mysterious element.

To keep the noxious witches' brew boiling, Marie had to stir constantly, using an iron bar taller than she was. After the pitchblende had been ground, they boiled it to separate liquid from solid materials. They discarded the liquid. The solids again were treated with chemicals to identify and separate each element. By grinding, mixing, dissolving, heating, filtering, distilling, and crystallizing – on and on, the ore sample got smaller and smaller, its radioactivity stronger and stronger. Summer notwithstanding, the fire under the cauldron burned fiercely. On rainy days, coughing and choking in the fumes, Marie had to transport the cauldron, back and forth: from outside, inside, outside, inside. When the entire mound of pitchblende was consumed, the yield was five or six grains of pure radium: one grain equals 60 mg; 1,000 mg equals one gram; 30 grams equals one ounce! Yet, Marie Curie remembers this arduous period as "... the best and happiest years of our life ... entirely consecrated to work."

The tiny Polish woman was surely exhausted from the long day of hauling a heavy cauldron back and forth, mixing poisonous compounds, endlessly stirring. After such a work-day, the family dinner had to be prepared, dishes washed, rooms swept, laundry scrubbed, a small child bathed and put to bed. (French husbands didn't do "female work" in those days!) Pierre's father "baby-sat" during the work-day. Later, he came again to their home and tended the sleeping infant. The couple would then sometimes walk back to the dark workplace. Marie described the spectral blue haloes in jars "arranged on tables and boards; from all sides we could see their slightly luminous silhouettes, and these gleamings, which seemed suspended in darkness, stirred us with ever new emotion and enchantment." What a picture! Two young people, standing in a dark shanty, surrounded by ghostly glows, silent and joyful!

Some evenings, they stayed home to write a complete summary of all their methods, all their experiments, all their findings. They soon learned that rays beamed by radium compounds, though invisible, were of great potency. (The notebooks of their daily chores and their inchworm program are still, 100 years later, so lethally radioactive, they are shelved in a lead-lined vault.) Another discovery: the rays didn't flame, but they burned! Might they "burn" away cancer cells? Tests at a nearby hospital confirmed the hypothesis. Skin cancers exposed to radiation were destroyed. When the burnt skin healed, it was healthy skin.

The news spread rapidly. "Curietherapy" became a buzz-word. They freely lent quantities to other scientists. They pursued this policy through their lives.

> From Norman Wyner's fine book, The Inventors – "The Curies could have made a fortune if they had patented their process for producing radium, but, poor as they were, they did not wish to take personal advantage of their discovery. They revealed their secrets to the world in the interests of humanity ..." Daughter Eve Curie quotes her parents: "If our discovery has a commercial future, that is an accident by which we must not profit ... radium is going to be of use in treating disease ... it seems impossible to take advantage ..."

Their devotion and endless hours of painstaking, often boring and often discouraging work finally yielded one of the great discoveries in the history of science: radium. In the words of Mme. Curie: "I submitted to a fractionalized crystallization two kilograms of purified radium-bearing barium chloride that had been extracted from half a ton of residues of uranium oxide ore." The procedure produced substances far more than 100,000 times the radioactivity of uranium.

That was written in 1899 – when, Marie admitted, "None of the new radioactive substances had yet been isolated. To believe in the possibility of their isolation amounts to admitting that they are new elements ... This tenacious property, which could not be destroyed by the great number of chemical reactions we carried out ... always followed the same path, and manifested itself with an intensity ..."

More work. Not until 1902 was Marie able to isolate a decigram of pure radium and determine its atomic weight. For this history-making achievement, she was awarded a gold medal. But gold medals do not put bread on the family dinner table. She took a job teaching physics to girls – the first woman to teach at that girls' school, of course.

Pierre addressed the Royal Institution in London (May, 1903); the all-male society had prohibited Marie from speaking. But she alone addressed a distinguished assemblage in the Sorbonne, reporting their newest "Researches on Radioactive Substances." It was the thesis paper for her doctorate. The degree "Doctor of Physical Science" was awarded, summa cum laude. Sweden's Royal Academy announced December 12, 1903, that the Nobel Prize for Physics had been awarded jointly to Marie and Pierre – and to Henri Becquerel. The Curie share was worth 70,000 francs; additionally, 60,000 francs went to Marie which she spent on installation of a bathroom (their first).

Pierre was promoted to full professor at the University of Paris, and was provided funds for assistants. The first hired was Marie, November 1, 1904. At last she was to be paid for her work. She took time out the next month to give birth, at 37, to Eve, her second daughter.

A normal, happy family life was of short duration. Pierre's health, never robust, had been deteriorating. His hands were disfigured by radium-burns. Probably there were internal damages, too. Possibly physical debility caused him to slip one day in a rainstorm, crossing a narrow street. He stumbled against a cart-horse thundering by, and fell under the rear wheel of the wagon. It killed him instantly. The date was April 19, 1906; Pierre was 47. From Eve Curie's biography of Madame Curie:

> "At the moment when the fame of the two scientists ... was spreading through the world, grief overtook Marie ... But in spite of distress and physical illness, she continued the work ... and brilliantly developed the science they had created together."

Pierre was buried in the family vault in a country village outside Paris. Condolences arrived from all over the world. Of consolation, however, there could be none. In Marie's diary, the entry read: "... Pierre, my Pierre, you are there, calm as a poor wounded man resting in sleep ... what a terrible shock ... (to) your poor head that I so often caressed in my two hands. I kissed your eyelids ... Everything is over ... it is the end of everything, everything, everything."

It wasn't. In the biography of her husband that Marie wrote, she quoted Pierre as having told her, early on, "Whatever happens, even if one should become like a body without a soul, still one must always work." Admitting "It is impossible for me to describe the meaning and depth of this turning point in my life ... I was unable to think of the future." But like so many prematurely widowed women, need played a part. She had two daughters under ten to rear.

Though a government pension for the widow was proposed, she refused it as "charity." She accepted, however, an offer by the Faculty of Science at the Sorbonne to replace Pierre. It was the first such appointment of a woman to a senior position in French education. She stood silent before the packed auditorium, November 5, 1906, took a deep breath and quietly, barely audibly, began her remarks with the last sentence that Pierre had uttered, in his last lecture, standing at that same podium. Newspapers trumpeted the event: "a victory for feminism. If a woman is allowed to teach advanced studies of both sexes ... the time is near when women will become human beings."

The salary was poor. The American millionaire Andrew Carnegie was so impressed by her courage in widowhood that he contributed $50,000 for scholarships at a new laboratory to be created, honoring both Curies.

A location was selected near the Pantheon. The director of the Pasteur Institute agreed to a collaboration. Biological research and radium-therapy would be centered in one building; radiation research in a second, headed by her. Together, they would be known as the Radium Institute of the University of Paris.

Meanwhile, Marie continued the search for pure radium. Four years after Pierre's death, she succeeded in isolating a tiny fleck of it by electrolysis with cathode of mercury. The achievement was celebrated by the award to her, November 7, 1911, of another Nobel Prize – this one "for her services to the advancement of chemistry by the discovery of the elements radium and polonium." No scientist, woman or man, had previously been honored by two Nobel awards – in two different scientific disciplines. (But two weeks later, her nomination to membership in the Academie Des Sciences was rejected! Prejudice against women abated slightly by 1933 when her daughter, Irene, was honored by a Nobel Prize for discovery of artificial radiation.)

July 31, 1914, the construction of the Radium Institute was completed. World War I broke out four days later. The slaughter of 10 million men began. Marie taught herself human anatomy, automobile anatomy, and how to drive. Often under fire, Marie and 18-year-old Irene (still a student at the Sorbonne) dashed about battlefields in an old automobile converted into a radiology laboratory. From the prototype, they created a fleet of mobile "Petit Curies" and 50 hastily-built facilities just behind the trenches. Men wounded in battle were rushed to the cameras. Within minutes after the soldier was felled, deep-lying bullets, shrapnel fragments, and broken bones were located on pictures, greatly facilitating treatment.

Some surgeons resisted the new-fangled technology. One, having probed into torn flesh for an embedded bullet, failed to find it. The schoolgirl Irene showed the doctor where he should have made his incision. Holding negatives up to a bare electric bulb, surgeons performed emergency operations just behind the trenches, without even waiting for printed plates. One million men were treated in these makeshift facilities.

After the Armistice, she resumed teaching radiology at the Sorbonne, by now a full professor at $160 a month. As this was her only income with which to support the family, she accepted $75 a week from the U.S. Army to train American officers in use of radiology equipment.

It was a busy routine, until 1920.

"Curietherapy" was the ONLY treatment yet discovered for cancer, the disease most women dreaded. Marie needed more radium – but by now, its cost had soared. An American magazine editor, Marie Meloney, smelled a fabulous publicity coup. She organized a fund-raising tour in America for the "Radium Woman." Marie was featured in dozens of publications. Parade routes were thronged by cheering crowds. She appeared at receptions, lunches, dinners, and met with government officials, including President Warren Harding. To encourage young women toward careers in science, she spoke at Smith, Mt. Holyoke, and Vassar Colleges.

Women rich and poor contributed; the campaign was a huge success, substantially over-subscribing the $100,000 goal. Marie returned to France with a fresh gram of radium.

A new career dawned – as a fund-raiser for French science. Beginning in 1921 with her first American tour, nearly to her death in 1934, she devoted much of her time and waning energy to soliciting money for research. Edmond de Rothschild contributed 10 million francs to finance scholarships for young science students; also 50 million for establishment of an Institute of Biology, Physics, and Chemistry. His and Rockefeller's donations created a Mathematics Institute. An Institute of Physical Chemistry was built adjacent to Marie's pavilion.

To keep the Western world focused on the needs of science, she attended innumerable international conferences. Among dozens of distinguished attendees, she was typically the only one without a mustache. On one such occasion, Albert Einstein observed "Marie Curie is, of all celebrated beings, the one whom fame has not corrupted."

As industrial uses of radium and its derivatives spread worldwide, many workers in laboratories and X-ray clinics started to complain of aching bones and extreme fatigue – symptoms of what later became known as radiation poisoning. (The very same disease that later killed thousands in Hiroshima and Nagasaki.) In the early 1920s, a New York dentist named Theodore Blum (who, with Leon Harris, fathered oral surgery) noticed telltale signs of cancer in the <u>mouths</u> of certain of his patients whose trade was to paint glow-in-the-dark numerals on the faces of precision instruments. Radium-based pigments created the glow. Some workers lip-licked their brush tips to sharpen the points. Blum published a warning: lesions could result in malignancies from radium poisoning.

The Curies and their colleagues in Paris did not take it seriously, though Marie had long been suffering. Early in their marriage, Pierre wrote to a friend that Marie was "always tired, without being exactly ill." During the four years of the radium-hunt, Marie had lost fourteen pounds. Cataracts developed in both of her eyes. She had four operations between 1923 and 1930, but her vision deteriorated steadily. Her kidneys were affected. She was afflicted with gallstones. Her bones became brittle. She entered a nursing home.

On July 4, 1934, aged only 67, Marie Sklodowska Curie died. She was buried beside her husband in the village cemetery. Sixty years later, the remains of both were exhumed and transported to the Pantheon.

But for fate
"Origin of the Species"
Might not have been authored by
CHARLES DARWIN

"England is tumbling toward anarchy, with countrywide unrest and riots; ... the fat, corrupt Established Church (is) ... a harlot in bed with the State ... to subjugate working people ... Britain now stands teetering on the brink of collapse."

Adrian Desmond and James Moore open their biography "Darwin" by thus describing the social and psychological climate prevalent in his lifetime.

The Industrial Revolution had propelled a vast emigration from rural areas into mushrooming, polluted smokestack cities, riddled with disease. There were many more brothels than churches. Workers were paid below-subsistence wages. The six-day week and twelve-hour day were norms. Child labor was endemic. Many farm women were bondaged slaves (well into the 20th century), with no say in their wages, no choice in where they worked or with whom they slept. Benjamin Disraeli, later to become Queen Victoria's Prime Minister, wrote "Infanticide is practiced as extensively and legally in England as (in India) on the banks of the Ganges."

The American Revolution of 1776, the French Revolutions of 1789 and 1830, and the brushfire uprisings in 1848 all over Europe had tolled a requiem for autocratic feudalism. Engels' and Marx's "Das Kapital" spread unrest and anti-clericism among the working classes. Dissent from religious dogmatism spawned a variety of "freethinker" sects.

A new political force of rich manufacturers and merchants was agitating to acquire power long monopolized by an oligarchy composed of the royal family, the nobility, the landed gentry, and the Church of England.

Both Oxford and Cambridge were controlled by the ecclesiastical hierarchy; clergy dominated faculty rosters. Prestigious appointments to university and museum posts favored blue blood over gray matter.

Notwithstanding, change was coming. Intellectuals and professionals eagerly read journals from France reporting new findings in science. Young men grew increasingly dissatisfied with domination of professions by the well-connected (often titled) old guard. Interest grew in study of anatomy, zoology, botany, a myriad of natural forms and forces.

Into this turbulence was born Charles Darwin in 1809. His mother died when he was eight. His father, the son of a famous physician, was preoccupied by his own very successful medical practice. Charles reached his majority as an

49

unambitious country gentleman who delighted in nothing so much as shooting birds and small animals. His older brother never bestirred himself to marry or to work. Charles might well have followed his example, but as there were no other sons in the family to maintain family fortunes, he was sent to medical school at Edinburgh University. Participating in operations on children, he was so appalled by the patient's agony – without anesthetics – that he dropped out of college. His disgusted father flayed: "You care for nothing but shooting, dogs, and rat-catching – and you will be a disgrace to yourself and the family."

Charles transferred to Cambridge in 1827, to prepare for a vicarage in the Church. In January, 1831, he earned the necessary Bachelor of Arts degree (without honors) but by then, his faith was not strong enough to take holy orders.

At Cambridge, his tutor, curate John Stevens Henslow, Professor of Botany, had interested him in plants and insects – a traditional hobby of provincial parsons. After his last spring term at St. John's in Cambridge, Charles accompanied the professor of geology, Adam Sedgwick, on a three-week hiking tour of Wales – and added fieldwork in that discipline to his enthusiasms. The summer ended; Charles returned home. Now what?

Having shied away from joining the church, drift seemed likely. His uncle, the famous Josiah Wedgwood II, was one of the richest industrialists in England (his porcelains are still treasured today). He wangled for Charles, late in the year, an unpaid post as the chronicler aboard the Royal Navy's "Beagle," about to sail on a two-year mapping expedition along the coast of South America. "The undertaking ... affords such an opportunity of seeing men and things as happens to few," he crowed, "not on the supposition of your being a finished Naturalist, but as amply qualified for collecting, observing and noting anything worthy to be noted ..." Because the appointment was without salary, Charles could have quit at any time. Nobody would have been surprised if he had.

The voyage started inauspiciously. Charles became seasick the day the ship left Plymouth, December 27, 1831; and he was seasick often thereafter, almost always when under sail. On one occasion, the ship was at sea 47 days. His first letters home were graphic: "I hate every wave of the ocean with a fervor." To his sister he confided "I loathe, I abhor the sea and all ships which sail on it." During the five-year voyage, the ship he abhorred was on the sea he loathed 533 days total, much of which he was wracked with nausea that confined him to hammock.

Like many sons of rich men, Darwin had not before been challenged by life. However, Edinburgh and Cambridge had taught him some useful disciplines. He took notes about weather, people and places. He was taught how to net, dissect, and preserve marine creatures. With no previous laboratory background, he learned use of the microscope.

Whenever the "Beagle" cast anchor, Darwin explored the flora, fauna, and geology inland. Also politics and economics. He often went on safari – forays not required by the loose terms of his posting, hence funded entirely from his pocket. He didn't gripe about hardships – rarely mentioned them in his journal. Many field trips lasted longer than three weeks; one actually took four months. He rode through swampy jungles; climbed the slopes and peaks of the Andes Mountains. He "lived off the land," subsisting on what he shot, and sheltering in a tent at night. At times, his group had to traverse so thick a primeval forest that horseback progress was a crawl over fallen trees webbing the ground to a height of 10 feet or more. He experienced an earthquake and volcanic eruptions; he parlayed with heavily-armed revolutionary insurgents; he faced off hostile natives.

Crossing a mountain range in Chile, he found marine fossil shells at 12,000 feet! He observed that certain species of vegetation developed differently facing east on one side of the peak than those on the other, facing west – the same plant! Did the variations imply different adaptations to different climatic, sun, and soil conditions? Later, he noted a similar anomaly in the lava-blanketed volcanic rock Galapagos Islands. Each of those bleak specks of land in the archipelago supported species related to those on neighboring ones – but developmentally different. Darwin charted 13 variations in the finch family, for instance. The birds on each island were different from those on nearby islands – but related to finches distant in the South American continent. More remarkable – even mysterious – in the process of evolutionary adaptation, they had acquired characteristics common to birds other than finches. Some searched for food like woodpeckers.

By this time, his letters home had a changed tune. He exulted in the "many magnificent and characteristic views … many and curious tribes of men" … the "fine opportunities for geology and for studying the infinite host of living beings … a prospect to keep up the most flagging spirit."

He collected insects and plants, vertebrates and invertebrates, prehistoric bones, stones, fossils, flatworms, plant parasites. He shipped to Henslow at Cambridge crates, casks, and barrels of preserved fish, skins, carcasses, embalmed rodents, seeds, rocks and feathers.

Every parcel was name-labeled with a full description of where the specimen had been found and when; color when collected; behavior if alive. At meetings of the Philosophical Society of Cambridge and the Geological Society of London, Henslow read excerpts from his letters. Long before Darwin returned to England, his reputation was established.

It might be assumed, from the foregoing, that Darwin had not yet focused on any one special interest. The assumption would be correct. It was his broad gauge of interests, his encyclopedic mobilization of facts, and his endless theorizing about them that eventually wove into the warp and woof of evolutionary theory.

The disparate strands of observations he noted did not cohere, but a pattern was taking shape. Toward the end of the voyage, he started a fresh notebook and wrote on the first page: "transmutation of Species ... greatly struck ... on character of S. American fossils – and species on Galapagos Archipelago ... these facts origin ... all my views."

In 1836, 27-year-old Darwin returned to Falmouth, England. He was surprised to find that he was renowned. Quickly, he was inducted into the most prestigious professional societies and clubs, invited to the homes of celebrities.

At that time, it was universally understood in Europe that mankind, created naked by God in the Garden of Eden, progressively grew more civilized, generation after generation. The well-born, affluent white male (English, of course) was the supreme example of the process. This comfortable assumption was severely shaken for Darwin by the paradoxes that he was studying. The "Beagle" years had broadened him greatly, but left him with the great question: was there a mechanism that governed them all? What he still had to learn had become "the making of my life," he wrote.

At the family home, he withdrew into a private world, analyzing minerals, barnacle shells, and bits of insect, plant and marine life. He continued to dissect, classify, sift and sort specimens. He also set to work writing a "Journal of Researches into the Geology and Natural History of the various countries visited by H.M.S. Beagle," and published the classic now known as "Voyage of the Beagle" in 1839.

The famous turning-point had been reached. "Notebook of Transmutation of Species" had fattened into a second, then a third volume. Darwin entered a notation dated 28 September, 1838 indicating that he had just read the "Essay on Population," by Rev. Thomas Malthus, an economist and mathematician. Its premise was simple: human fertility was such that, unimpeded, "population ... (will) increase at geometrical ratio ..." Food supplies, at best, could only increase arithmetically. The planet would become unlivable. What deterred the catastrophe were natural disasters such as earthquakes, hurricanes, forest fires, floods, famine, disease. And human competition for food, for sex, wars for power and living space. The "Malthusian Doctrine" was instantly recognized by Darwin as applicable to all living species, whether they flew, swam, crawled, or swung through trees. Life at all levels was a battle. Only the dominant specimens in any, in every, species survived. They, only they, lived to procreate and deliver the next generation – with superior genes inherited from successful progenitors. He remembered a lecture in Edinburgh at which his good friend, the iconoclast Robert Grant, had described the transmutation metamorphosis as "evolution," (Grant was the very first to use that word.)

In a letter on April 6, 1859, he wrote "I came to the conclusion that selection was the principle of change from ... reading Malthus (and) I saw at once how to apply this principle."

Soon after the epochal discovery in Malthus, Charles proposed marriage to his childhood friend and cousin, Emma Wedgwood. They wed in January, 1839; rented a large house in London; hired a butler and domestic staff. Children were born.

Depressed by the tumult of a growing household and London's teeming streets, he bought a property in Downe, rural Kent, blessed with fresh air and distant views. Though only a score of miles from London, it was remote from a rail-station. There the family lived quietly for decades, trying a variety of "cures" for Charles' increasing health problems. He had just enough strength to average two hours of work, daily. His physician described Darwin's regimen: brood through a sleepless night until the morning, when he would "expose the night's hypothesis to realistic examination and (then) complete the day's work." He suffered from a myriad of ailments that affected concentration. His writing was sometimes interrupted by long interludes at sanatoria.

For change of pace, he rambled the countryside, interviewing sheep and cattle breeders, gardeners and naturalists who had practical experience at measures to improve their species. He corresponded with zookeepers. He read voluminously. Years passed, but he really was in no hurry. Conservative in all matters, he had no appetite for flying in the face of Authority by expounding a natural - selection theory; strict laws defined blasphemy. A clandestine copy of Sir William Lawrence's 1818 long-banned lectures on natural history sat on Darwin's desk. It was a precursor of the new physiology in science – and a reminder of what could happen "swimming against the tide." Darwin – like all scientists of his generation – knew that Lawrence was forced to disavow his beliefs. Darwin had no appetite for such martyrdom. At times, his confusion was such that he thought not to publish "Origin of the Species" at all. Though himself not a churchgoer, he respected the piety of his wife, who fervently believed in Revelation. He instructed her to engage editors after his death, specifying a fee of 400 pounds to prepare only a summary of his notes for publication.

The postman's knock on his door, Friday, June 18, 1858 was a fateful intervention. He delivered a package, sent from Asia by Alfred Russel Wallace, a free-lance naturalist. He had corresponded with Darwin, and, from time to time, had shipped to him specimens for which he was paid. He was respected for his earlier discoveries in South America, which had led to publication, in 1853, of "Narrative of Travels on the Amazon and Rio Negro." In 1855, in Borneo, he started to write "On the Law Which Has Regulated the Introduction of New Species." It was imperfect, incomplete, until – sleepless and shaking with malaria fever in February 1858 – the Malthusian Doctrine flashed into his mind. He wrote a 20-page summary of a theory of natural selection. He noted the struggle for existence, the geometric rate of increase in animal populations, the limited food supplies, and concluded that "those that prolong their existence can only be the most perfect in health and vigor."

In the June 18 package, he enclosed that 20-page manuscript. Without the range of examples that Darwin had laboriously assembled, and lacking the inductive rigor of Darwin's reasoning, Wallace had clearly outlined the main ideas of transmutation. "He could not have made a better short abstract," wailed Darwin. "Even his terms now stand as heads of my chapters!"

Biographer William Irvine, in his immensely readable "Apes, Angels and Victorians" says "What Darwin had puzzled and wondered and worried and slaved over with infinite anxiety for two decades, Wallace had (arrived at) the same results." The Wallace book, if published first, would gravely damage readership of the Darwin book, when it was later published. If it were later published.

What to do? It was not in the character of a Victorian gentlemen to act dishonorably. "I would far rather burn my whole book than that Wallace or any other man should think that I have behaved in a paltry spirit."

A few days earlier, Darwin's retarded infant son had died of scarlet fever – one of six children in the village to be carried off by the disease. Darwin feared the epidemic would spread to his other children. Depressed and demoralized, he lacked clarity of mind to determine a course of action. Close friends were asked for advice. Joseph Dalton Hooker, scion of a titled family, was among the leading botanists in England; and Sir Charles Lyell's three-volume "Principles of Geology" had been Darwin's bible during the "Beagle" voyage.

Therein, Charles had read, years before, "In the universal struggle for existence, the right of the strongest eventually prevails; and the strength and durability of a race depends mainly on its prolificness ..." (Dr. Sandra Herbert, of the Smithsonian Institution in Washington, in noting the connection, concludes that Lyell "impelled Darwin to apply what he knew about the struggle at the species level to the individual level, seeing that survival at the species level was the record of evolution, and survival at the individual level its propulsion.")

The two eminent scientists devised a strategy that effectively finessed the dilemma. Thursday, July 1, before a hand-picked group of scientists at the Linnean Society in London, they read aloud from Darwin's unpublished texts those paragraphs that developed the evolution theory. Then they read the Wallace paper "On the Tendency of Varieties to Depart Indefinitely from the Original Type." They thus "put on record" simultaneity of the rivals' authorship. Nobody in the audience recognized that history was being made.

Wallace wrote, years later, that the "theory of Natural Selection I shall always maintain (is) yours and yours only. You had worked it out in details I had never thought of, years before I had a ray of light on the subject, and my paper would never have convinced anybody or been noticed as more than an ingenious speculation ..."

Wallace's 20 pages galvanized Darwin into action. Within weeks, "Wallace's impetus seems to have set Darwin going in earnest ...," Thomas Henry Huxley, a close friend, said. "I look forward to a great revolution being effected."

Darwin was a slow writer, somewhat "Teutonic" in his determination to achieve clarity of meaning. But he was a keen observer, a careful note-taker, eloquent. On a mountain foray, he survived a gale that "lasted for more than a day; the men began to lose all their strength and the mules would not move onwards." Not at all interested in economics or social justice, he nevertheless provided detailed information about exploitation of expatriate Welsh miners in Chile. The shaft he visited was 450 feet deep; the average load a man had to carry up a tree-trunk "ladder," 12 times a day, he wrote, weighed 200 pounds. (Darwin actually carried one up to the scale.) He explained that miners were not permitted to stop for breath while ascending. Their daily diet changeless: "sixteen figs and two small loaves of bread ... boiled beans ... roasted wheatgrain."

In "Origin of the Species," he hoped to avoid controversy by never mentioning that man was just another animal subject to the laws of natural selection. The ending of his 150,000-word-tome tries to camouflage the implications of evolution:

> "Thus from the war of nature, from famine and death, the most exalted object which we are capable of conceiving, namely, the production of the higher animals, directly follows. There is grandeur in this view of life, with its several powers, having been originally breathed into a few forms or into one; and that, whilst this planet has gone cycling on according to the fixed law of gravity, from so simple a beginning, endless forms most beautiful and most wonderful have been, and are being evolved."

"Origin of the Species" appeared in print November 22, 1859. the first edition of 1,250 copies sold out immediately, and hundreds of customers remained on the waiting list. Revised editions and foreign translations of the book continued to come off the presses as long as Darwin lived. He received two-thirds of the net trade income and became a very, very rich man. Wallace continued to write and publish, but remained poor. (In January, 1881, a petition signed by Darwin and friends triggered action by Prime Minister Gladstone. Thereafter, Wallace received an annual pension of 200 pounds sterling from the Treasury.)

The Darwin book fell upon Britain like a bolt of lightning. Rev. Adam Sedgwick, one of Darwin's former tutors at Cambridge, voiced dissent and outrage, charging that Darwin's theory "spawned a hideous monster; that it would be merciful to crush the head of the filthy abortion and put an end to its crawlings."

Liberals hailed the author as "the greatest revolutionist ... of this century." Even Rev. Charles Kingsley – famous for his brand of "muscular Christianity" – found the Darwinian theory "just as noble a conception of Deity, to believe that He created primal forms capable of self-development ..."

Not so open to new thinking, traditionalists condemned "Origin" as sinful, heaped invective, abuse, and derision. A heavy blow was struck by Louis Agassiz, at Harvard, the world-famous authority in disciplines as diverse as geology and zoology. He was dogmatic that all species evolved into higher forms under the influence of divine plans – the "development hypothesis" as it was widely known. He ridiculed those who espoused Darwin's "evolution" as "alchemists." (Agassiz died disappointed. His own son, Alex, a marine biologist, became a believer in natural selection. As a final irony, the marble statue of Agassiz outside Stanford University's Zoology Building toppled from its pedestal during the 1906 earthquake. It was unbroken, but a celestial joker had guided it to rest upside down; its head broke through the concrete to become buried in the sand.

His eminence, Cardinal Newman, suggested that evidence was not the best test of truth. Thomas Henry Huxley, Darwin's close friend, shot back: people "must be prepared to choose between the trustworthiness of scientific method and the trustworthiness of that which the Church declares to be Divine Authority." Alvar Ellegard's "Darwin and the General Reader" (subtitled "The Reception of Darwin's Theory of Evolution in the British Periodical Press"), reveals that confusion typified the reactions of critics and ordinary citizens.

All of which led to a still-famous confrontation between science and tradition. In Oxford, on June 30, 1860, the 30th annual conference of the British Association for the Advancement of Science drew a capacity audience, mostly clerics. Not including Darwin, who was shy by nature and detested controversy, sent regrets, alleging illness. Actually, he was busy on his next book "On the Various Contrivances by Which British and Foreign Orchids Are Fertilized by Insects, and on the Good Effects of Intercrossing." Published in 1862, this study preceded Gregor Mendel's seminal research into pollination biology.

Darwin's proxy at the Oxford showdown was Huxley. His opponent was ornithologist and mathematician Samuel Wilberforce, Lord Bishop of Oxford. The prince of the church evoked cheers and laughter by asking his young adversary – barely past 30 – whether it was through grandfather or grandmother that he claimed descent from a monkey? Huxley replied that he would "rather have a miserable ape for a grandfather" than (nodding toward the splendidly-robed clergyman) "a man highly endowed by nature and possessed of great means and influence for the mere purpose of introducing ridicule into a grave scientific discussion." The audience responded tumultuously. Women fainted.

Admiral Robert Fitzroy, skipper of the "Beagle" and Darwin's companion during the five-year voyage, rose to his feet, waving a Bible over his head, shouting that it was the source of all truth. The debate is now memorialized on a plaque at the entrance to the Science Museum. Oxford residents still chuckle when they relate the story, referring to the hapless bishop as "Soapy Sam." Wilberforce later wrote a lengthy critique of "Origin" which appeared in the prestigious Quarterly Review, exposing what he saw as a fatal flaw – failure to deal with the evolution of man in the theory of natural selection.

In his 1959 book "Asa Gray," A. Hunter Dupree underscores the significance of this encounter. The Oxford clash "transformed the whole nature of the conflict over Darwin's theory. Before ... it had been the old battle within science between idealism and empiricism ... The Bishop of Oxford ... made the conflict appear in the eyes of the world as one between religion and science."

Early in his book, Darwin set forth his position with eloquent simplicity:

> "Let it be borne in mind how infinitely complex and close fitting are the mutual relations of all organic beings to each other and to their physical conditions of life; and consequently what infinitely varied diversities of structure might be of use to each being under changing conditions of life ... If such (variations) do occur, can we doubt that individuals having any advantage, however slight, over others, would have the best chance of surviving and procreating their kind? ... This preservation of favourable individual differences and variations, and the destruction of those which are injurious, I have called Natural Selection, or the Survival of the Fittest."

He then, in dealing with religious orthodoxy:

> The more I study nature, the more I become impressed with ever-increasing force, that the contrivances and beautiful adaptations slowly acquired through each part occasionally varying in a slight degree but in many ways, with the preservation of those variations which were beneficial to the organism under complex and ever-varying conditions of life, transcend in an incomparable manner the contrivances and adaptations which the most fertile imagination of man could invent."

In the first edition of the book, the phrase "this view of life, with its several powers, having been initially breathed into a few forms or into one" conspicuously omits any source for the breathing into life. Within weeks, the second edition inserts "by the Creator" preceding "into a few forms or into one."

Most senior-age scientists – even with vested interests in orthodoxy – recognized that a line had been scribed in the sand. Wallace wrote "The history of science hardly presents so striking an instance of youthfulness of mind ... as is shown by ... abandonment of opinions so long held (by them) and so powerfully advocated."

The sound logic of Darwin's arguments were, within a decade, accepted by most educated people, as well as scientists. So in "Descent of Man and Selection in Relation to Sex," Darwin, twelve years later in 1871, boldly ventured farther – into sexual selection and ape ancestors. He confidently discussed the conflict with religion and morality.

By then, Darwin was probably the most famous scientist in the world. The Kaiser of Prussia and the Russian Czar decorated the English naturalist. But when a Cabinet member, Lord Derby, wrote to Prime Minister Disraeli in 1872, suggesting that Darwin, whose "reputation is European," should also be recognized with a baronetcy at home, "Dizzy" vetoed the idea, wary of offending church factions. This was the second time he was denied a knighthood. Prince Albert had agreed to the honor in 1858; it was vetoed by Queen Victoria on the urging of Bishop Wilberforce and a group of clerical advisors.

When Darwin died, April 19, 1882, and it was planned to bury him in Westminster Abbey, many Anglican prelates bemoaned the blasphemy from their pulpits. Among religious fundamentalists, "Creationism" is still alive, the theory now often described as "intelligent design." A Gallup Poll revealed that nearly half of Americans believe that God created mankind within the past 10,000 years.

"The only rational course for those who (seek) the attainment of truth (is) to accept "Darwinism" as a working hypothesis," Huxley said in 1887, "and see what could be made of it." In a recent "Commentary" essay, David Berlinski, points out that gaps in the theory still invite study. "Most species enter the evolutionary order fully formed and then depart unchanged."

One of the world's leading Darwinians, Oxford's Richard Dawkins, has called for evolution of Darwin's theory itself. In "The Selfish Gene," (1976) he coined a word – "meme" – terming it a "social parallel to biology's gene." He says that "Selection favors memes that exploit their cultural environment to their own advantage ... (and) live on intact, long after —- genes have dissolved in the common pool ... Cultural transmission is analogous to genetic transmission —- it can give rise to a form of evolution." Ian Stewart, Professor of Biodiversity at Warwick University in England, argues in his book, "Life's Other Secret" that spontaneous self-organization of physical material (environmental phenomena) is as essential to life as the genetic code that underlies it. In essence, he concludes that nurture is as structured and vital to survival as nature.

Inadequately weighed also is Gregor Mendel's axiom that for each local adaptation resulting in a "higher" form, another equally successful local adaptation can evolve by "degeneration" of morphology or behavior. Equally worthy of thought is the effect on genetics of modern medicine, which keeps alive the weakest children and attenuates the lives of those with diabetes, poor eyesight, other once-certain victims of natural selection.

In "The Possible and the Actual," Nobel laureate Francois Jacob challenges: "the theory of evolution provides a framework without which there is little hope to understand where we come from and why we are as we are ..." Should we add: "where we are going?"

Advocate for those Who
had no other advocate
DOROTHEA DIX

Successful in three different careers before she was 40, early in nineteenth century America, Dorothea Dix did not know what she wanted to do with the rest of her life. She had headed a school. She had authored highly popular books. Such a brilliant career was rare for a single woman in nineteenth-century America. Yet she was dissatisfied. There had to be something more to challenge her intelligence and engage her spiritual needs. Then she was asked to teach inmates in a prison. A chain of events and a trail of service followed that has few equals in American history.

Her grandfather had been a self-made millionaire, murdered at the height of his success. Her father, Joseph, when assigned to manage some of the family businesses, managed them into worthlessness. After which, he couldn't hold a regular job, couldn't stay sober, couldn't provide what every child needs, domestic stability. Then he heard "the call" and became a self-appointed minister in religious fundamentalism that was sweeping through New England. As usual with deviant sects, it was held in contempt by "quality" folk, most of whom were high church. Even kindly Ralph Waldo Emerson, normally the most tolerant of men, ridiculed their worship practices, "jumping about on all fours."

Ridicule is discrimination practiced by polite folks. Lower-class bigots resorted, rather, to violence. Often, Joseph and his family were driven out of town. Dorothea's earliest years were marked by a succession of such violent rejections. Her father's improvidence forced the family to move often, each poor home meaner than the one before.

Loveless and so often humiliated at a vulnerable age, 13-year-old Dorothea begged sanctuary with her widowed grandmother. The old lady's home in Boston was a memorial to her late husband, stuffed with souvenirs of his travels. Oppressed in that somber environment, Dorothea was testy and stubborn. In less than a year, caretaker responsibility was passed to an aunt who was already fostering Dorothea's two much-younger brothers, abandoned by their feckless parents.

The girl assumed authority over the boys and quickly gained self-confidence and a degree of maturity. She found distasteful the frivolous world of females her age whose social life consisted of parties, teas, dances, and gossip before "taking their place in society." The teen-ager wrote "I have little taste for fashionable dissipation ... look with little envy on those who find their enjoyment solely in such transitory delights." Instead, she organized her first school, accepting even infants for day care. During the years when other girls were trained to become social ornaments, homemakers and mothers, Dorothea acquired administrative and teaching skills.

She wrote "My young brothers rest themselves on me ... the feeble frame of my aged grandmother seeks for support ... I have none to cling to, none that I love in this world ..." She yearned to become a "fit companion" of a "virtuous great man" who was "perfect in the way of righteousness."

Puritan religiosity in New England was gradually warming into coexistence between revelation and reason. Congregationalism and then Unitarianism spread among some of the upper-class churches of Boston. Irish and German women immigrants were relegated to a "below the stairs" world; menfolk to factories and mines. In the face of the deepening gap between classes, clergymen sermonized their flocks not just to look toward an after-life, but to serve God "here and now" by attending to the community's social needs. Dorothea appealed to her frosty grandmother, an Episcopalian of course, to support a charity school for poor children in the Dix mansion: "God has placed us here to serve himself in serving his children of earth," she lectured. The old woman was unmoved, flatly declined.

Dorothea attended Unitarian services frequently, taking notes during the sermons, writing summaries. She scoured residential neighborhoods, collecting books, clothing, and shoes to contribute to ministers who recycled them to the poor. Her uncle, Thaddeus Mason Harris, had been graduated from the new Divinity School at Harvard, and been assigned a new pulpit in Dorchester. Unconstrained by embedded traditions, he gently led the congregation into Unitarianism. In her twenties, Dorothea was a regular in his group visits to the Boston Female Asylum and the jail in East Cambridge.

Another relative, cousin Thaddeus William Harris, had been graduated in medicine from Harvard. He introduced Dorothea, an avid reader, to the natural sciences. Other than from the two Harris relatives, she never had normal schooling, but Dorothea acquired the smattering of a broad Harvard education. By the time she was in her early twenties, Dorothea's mind was churning with a mixture of disparate interests: religious extremism; theological revisionism and rationalism; secular literature and readings in the natural sciences. A rich goulash it would have been for a man to function in the male world. But of what relevance to a young female? How did her education prepare this young woman for the Proper Bostonian marriage market? Dorothea was in the bloom of her youth, tall, graceful, dressed fashionably. She had self-assurance and upper-class manners. But she was humorless and "good works" were the highest goals of her life. It was clear to any potential suitor that she would not be "a woman (who) loses her independence forever in the bonds of matrimony," as was expected in America, the French social historian, de Tocqueville observed.

One "candidate" after another came and went. These repeated rejections were hurtful, of course. How deep the wounds? Decades after they had drifted apart, a former male friend received a "congratulations on your marriage" letter from Dorothea, asking for a lock of his hair. Anticipating a meeting with Lafayette, the celebrated hero of the Revolution, Dorothea gushed to a friend, "I long ... to be in the room with him ... I hardly would trust myself ..."

Clearly, teaching pupils and attending church were not enough to satisfy Dorothea's energies. Her uncle Harris had published in 1803 a four-volume "Minor Encyclopedia," crammed with knowledge — geographic, historical, scientific. Dorothea, with a similar idea in mind, compiled "Conversations on Common Things; or, Guide to Knowledge" – for children. Its format was unusual, a dialogue between a hyper-curious young girl and her mother. Dipping into her hodgepodge of knowledge, Dorothea had the precocious child ask questions about bizarre subjects — with no rhyme to the selection process except that they were all questions to which she, Dorothea, could research answers. The book implied equal intellectual capacity in male and female children. It became a best-seller.

Flushed with her success, Dorothea at 23 decided to transform religious education into a reader-friendly format for children. She assembled Christian hymns and interspersed them with excerpts from Buddhist and Hindu teachings. She sandwiched poems and psalms which were her own creations. Another big success. Over the following years, she penned three more books, also of devotional and inspirational content, also money-makers. By her mid-twenties, she felt confident enough to write fiction. In "Ten Stories for Children," she dredged memories of her own childhood, reworked them into morality tales. All cast a baleful light upon parental irresponsibility, but invariably the family reached a happy ending through their children's grit. Royalties thereafter contributed to her lifelong independence, rare for a single woman.

Dorothea's remarkable amalgam of being a mildly celebrated author, a moral exemplar, and a respected schoolteacher was irresistible to Unitarian ministers with large families. A succession of them employed her as governess for their children in Boston. Then she was called to Philadelphia. The progressive Quaker-dominated city government had just sponsored a survey of workhouses and almshouses. Dorothea on her own initiative, and on her own time, undertook to study Philadelphia's schools, which included a rarity, an institution for the "deaf and dumb." These children had been rescued by the Friends from neglect in environments where they were considered to be God's mistakes. The experience sank deep into Dorothea's psyche.

She possessed great energy, high idealism, and boundless ambition; however, none of these qualities had a natural outlet which could be employed in a conventional profession or vocation, women such as Dorothea. Back in Boston at 27, and severely depressed, she accepted the invitation of a famous Unitarian minister whom she greatly admired, Dr. William Ellery Channing, to join his family as a teacher for his children during an extended stay on the Caribbean island of St. Croix. She taught arithmetic, spelling, grammar, science. ("Strict and inflexible," her pupil, Mary Channing later testified.) She was enraptured by the flora and fauna of the tropics. She wrote her next book, "The Garland of Flora," which displayed not only vast botanical and literary erudition but poetic writing of high quality. She described exotic native blooms, incorporating passages from English and French poetry. She used references from Virgil, Cicero, Plutarch, Chaucer, Shakespeare, Jonson, Milton, Dryden, and others. She quoted from Johnson's "Rasselas" and Keats' "Endymion."

By the time the St.Croix interlude ended, Dorothea's health was somewhat improved, but not her melancholy. Diary entries at this time remark on the oppression of "faithful conscience" that "brings my secret sins to view." Sins? Biographers assume she still suffered from the guilt of rejecting her parents. Her father had died, Dorothea never wished to see her mother, though she supported her. She referred to her grandmother as "Medusa." Still influenced by Wesleyan Methodism, it would seem she accepted its dictum "it is unholy to be happy." Gloom and hopelessness were seen as necessary steps toward salvation.

Concerned by her deepening depression, friends in the Unitarian ministry arranged for her to visit England, hoping that the sea voyage would be of benefit. At 30, a depressing age for a spinster, Dorothea set sail. The trip added physical fatigue to her condition. By the time the tiny ship reached dock in Liverpool, Dorothea collapsed with bronchitis and lung-congestion. Her innkeeper, shown a letter of introduction from Dorothea's former employer, Channing, to an old friend in England, William Rathbone, reported to Rathbone. He immediately came for her and installed her in his spacious manor.

William had just been elected mayor of Liverpool on a social-reform platform. An advocate of public schooling, he was also an "overseer" of the lunatic asylum, one of the first to be separate from prison, workhouse, or almshouse. The Rathbone home was a magnet for visitors with liberal political and social views. Many were men of prominence, soon to assume control of the national government. Ideas and idealism were the mainstays of conversation. Dorothea breathed in the liberal, socially-conscious philosophy they represented.

She also soaked up information about England's progress in treatment of mental illness. Removal of persons deemed to be "insane" from habitation among felons was thought to be a vital first step toward rehabilitation. William Tuke, a friend of Rathbone's father, had, in 1796, established the "York Retreat," to be "a place in which the unhappy might obtain refuge — a quiet haven ... (to) find a means of reparation or safety." Therapy was in the tradition of the gentleness and respect of the Society of Friends – who believed that all, however eccentric, possess the spirit of God. In York Retreat, chains were not used, nor straitjackets, nor restraint of any kind save in the most extreme need. Combining religious belief with social service was balm for Dorothea's melancholy — and inspiration for many of the ideas that later animated her work.

After months of rest in the Rathbone mansion, her health improved sufficiently to plan return to the United States — with a clear idea of the role she would pursue. She wrote: "The suffering — to be comforted, — the wandering led home ... the indolent roused; the over-excited restrained."

Back in Boston, in 1841, a Unitarian minister, James Nichols, asked her to teach class in the East Cambridge jail. There, she was shocked so see that cells for the insane had no heat. Why? she asked. The answer: lunatics did not know the difference between cold and hot, so would hurt themselves if they had heat — or would set the building afire. Unable to convince the officials otherwise, she petitioned the court to require that heat must be provided. It was so ordered. Nichols later wrote: "Thus was her great work commenced."

She made the circuit of the city's jails and poorhouses. In each case, she specified measures of reform — and forced them through. She then "took her show on the road." Starting with the almshouse in Lowell, she toured the length and breadth of Massachusetts, making copious notes as she inspected each facility. In her boardinghouse — a different one nearly every night — she fleshed out the day's notes. When she was finished, she wrote a lengthy document reporting her eye-witness findings.

She identified herself as "as an advocate of helpless, forgotten insane ... men and women ... sunk to a condition ... (of) real horror" ... reporting "cold, severe facts ... from them you may feel more deeply the imperative obligation which lies upon you." She provided names and locations where she had seen insane women, cramped into straw-lined packing crates, "entirely naked — nor was there any garment in the place which could be used to cover them." Elsewhere, in a cavernous hall, she had observed "a great, monstrous, horrid company ... crying, shouting, laughing, screaming, moaning ... rolling on the floor ... and rousing each other to higher and higher exasperation."

The indigent insane, she lectured, are done great injury "by confining them in Jails and Houses of Correction .. (it) retards the recovery of the few ... and may render permanently insane (those) who, under other circumstances might have been restored to their right mind ..." She assaulted Victorian morality by painting a graphic picture of feeble-minded females, violated by other inmates and prison "keepers," piteously confused by newborn infants in their arms. She described "insane persons confined ... in cages, closets, stalls, pens ... Chained naked, beaten with rods, and lashed into obedience."

By citing shortcomings in local institutions, she established pertinence to every reader in every town. Overall, she charged the State of Massachusetts with breaching the Constitution by consigning the insane and dull-witted — though they had never been convicted of a felony — to "cruel and unusual punishment." She employed the florid rhetoric she had learned at backwoods revival-meetings in her childhood. She then leavened her diatribe with Christian hope, promising wonders when community leaders devoted themselves to social responsibilities.

She scorned "do-gooders," who funded missionaries to establish Christian outposts abroad, while their neighbors were in want at home.

She badgered the legislators, who were reluctant to spend the money building the type of an institution never before needed. Day and night, she called on them at home and office, wrote endless letters. She plagued them to "confront ... ghastly facts and inescapable conclusions," as David Gollaher puts it in his splendid biography, "Voice for the Mad." The bill for construction of an "Asylum for Incurables" was passed and signed into law by the governor on the last day of the 1843 legislative session.

This was the turning point in Dorothea Dix's life. Barred by her sex from entry into politics or any other traditionally male profession, she applied her talents, idealism, and energy into creating a new one — advocate for "God's mistakes." She immersed herself, day after day, in visiting institutions, seeing and hearing the problems of housing, feeding, treating the inmates — and devising measures to benefit those who hadn't the wits to care for themselves. What appalled others didn't faze her; she had first-hand acquaintance with misery as a child in New England's backwoods.

In the years that followed her Massachusetts triumph, she invaded — uninvited — state after state: New York, Rhode Island, New Jersey, Pennsylvania, Maryland, Ohio, Virginia, Carolina, Tennessee, Indiana, Illinois, Alabama, Louisiana, Mississippi, Florida, Georgia. Her writing gradually abandoned evangelistic fervor and strident accusations of neglect, in favor of calm essays with sensible proposals for reform. Only occasionally, when resistance was stubborn, would she get tough: "The establishment of hospitals for the insane has, within the last century, become so general among all civilized and christianized nations, that the neglect of this duty seems to involve aggravated culpability."

Every good salesperson knows to focus on what is important to listeners, and to speak in terms meaningful to them. Every good politician and journalist knows that all business is local, the issues that matter to each voting constituency are local concerns. Dorothea tailored her arguments to apply with pertinence in each area she toured. That was the easy part. What she found far more difficult to deal with was the "bribery and corruption ... unmasked and unabashed." She despised the need to pander to the personal interests, egos, and prejudices of politicians. But she forced herself to; she used whatever tactics would be appropriate. On one occasion, she nursed the sick wife of an important politician, to win his vote.

The country was growing; immigration was swelling; populations were moving into new territories, bringing unprecedented needs and problems. Here and there, ethnic riots broke out. Legislators had issues to deal with – if they hoped to be re-elected — more important than Dorothea's. Often, the problem was insufficient money. She devised the attractive blandishment of proposing a building program that would "pay for itself." She promised that if a minimum of 100 acres was dedicated, the inmates of the asylum, under supervision, would till kitchen gardens. They would greatly benefit from the exercise when released from their cells; and their surplus harvests, sold outside, would produce revenue.

The Eastern seaboard of the United States was serviced, in Dorothea's day, by few railroads. Except for when stern-wheelers on the great mid-western rivers could call on a target area, nearly all of her travel, visiting every city, town and village with an institution of confinement, was by barge, horse-buggy, and stagecoach, bouncing along rutted roads. "Considering it is almost unknown ... for one lady to make a journey by herself," she wrote to a friend, I have made my way quite successfully." Rivers, swamps, dark forests — nothing daunted her. However long the trip, however difficult, she would reach her destination.

In each location, she would first identify potential allies as a first step. Using them for introductions, she would interview leaders of every group that might take a position on her petition for a new asylum. All were men, of course and gender gallantry required that they then accord courtesy to such a fine lady with such impressive connections. She was especially effective in southern states, where her elegant manners were warmly appreciated. She was slim, expensively garbed, brainy. With some men, she flirted. She became friendly with governors and the leaders of factions in state assembles. She became so skilled at her work that in some communities, movements sprang up to dedicate monuments to her (which she declined); in some, the institution she had brought into being was named after her; some dedicated wings of their buildings to her. The Pennsylvania State Lunatic Hospital in Harrisburg has, under its cornerstone, a copy of Dorothea's original petition.

Coalitions that included doctors, lawyers, judges, teachers, ministers, justices of the peace, begged her to come to their community and organize their efforts. She was like a force of nature — in one case, she railroaded a bill through a reluctant legislature within four weeks of her arrival. Political scientist Francis Lieber described her as having the "firmness, courage and persevering strength of a male mind, ... unite(d with) the advantage of a woman."

A by-product that emerged from the vast amount of research and field-work she undertook in behalf of lunatics was authorship of a lengthy "Remarks of Prisons and Prison Discipline in the United States," published in 1845. For the first time, she broadened her survey from parochial to national.

After five years of almost incessant travel, she settled in Washington, D.C. in 1847. She had developed an appetite for maneuvering within the corridors of power; Washington was where they centered.

Vast new territories had expanded the country's boundaries, appropriated as a result of the Mexican War. Millions of acres in the South and the West became available. To Dorothea, these virgin areas represented an extraordinary opportunity. She was familiar with federalism in England, where central government in London was involved in the affairs of every province. Why shouldn't the same principle apply in America, too? Shouldn't the national government use national resources to care for its disadvantaged citizens? Not a bizarre notion, considering that the Constitution so directed.

On 17 June, 1848, she floated a trial balloon, in a pamphlet proposing "A grant of land for the relief and support of the indigent curable and incurable insane in the United States." Then set about weaving a fabric of Congressional support, starting with contacts among legislators she had befriended during her campaigns in their states. Congressman Horace Mann and Senator Charles Sumner, both from Massachusetts, were prominent among them — but there were many others from every state. Never identified as a political partisan, she moved easily across party lines. Many solons were enticed by the seemingly irresistible logic of her proposal: to auction federal land (for which no use was then envisioned); and then allocate some of the proceeds to build institutions for the insane, for the blind, deaf, dumb. By buttonholing congressmen and senators in the halls of the capitol, by visiting them in their offices and homes, Dorothea recruited backers one by one. President Polk gave his blessing, Dorothea having been introduced to him by friends in his home state.

Many, however, did not react to her idea with enthusiasm. The greatest political crisis in U.S. history was taking shape — conflict between North and South. A new president, Zachary Taylor, was inaugurated in 1849 — and he was a slave-owner. The slavery question was at the heart of the social structure and political economy of the South. In addition, "states rights" were an issue as old as the union. Only a half-century before, that principle had agonizingly delayed enactment of the federal Constitution. Thomas Jefferson's document was only approved by devising the most delicate of compromises. Nothing since 1791 had softened Southerners' determination to preserve their independence. Intrusion of federal money and administrators — hence federal power – into affairs that had hitherto been considered "states rights" was a mortal threat.

Dorothea thought to outflank opposition by dangling a compromise — the new institutions would be built with federal money, but would then be turned over to the states for administration, and become their properties. Some Southern legislators took the bait, others didn't, opposed to any extension of federal power in their domain.

Dorothea had known every kind of opposition during her five years of campaigns, state by state. Now began five years of constant politicking, stretching over a series of legislative sessions. Five years of wooing diverse personalities with diverse interests. Five years of horse-trading, deal-making, moral exhorting, ego-massaging. She proved to be so good at this work that one smitten senator suggested a permanent conjugal arrangement — though he was married.

Dorothea was granted a private alcove in the Congressional library as an "office," to confer with a parade of visitors, 9 a.m. to 3 p.m. daily. When Zachary Taylor died suddenly in 1850, she focused upon the new president, Millard Fillmore, and he became a close friend, eventually a confidante.

Dorothea's painstaking efforts brought her cause close to victory in 1851. Her Bill was approved by a majority exceeding two to one in the Senate. On the eve of calling a vote in the House of Representatives, the Bill was sidetracked in a raging debate over the flouting of the Fugitive Slave Law by Massachusetts Abolitionists. Introduction of a flood of pro-slavery and anti-slavery measures swamped the House. The legislative session ended, and with it, the prospect of victory in 1851.

In 1852 she used her close friendship with President Fillmore to slip through a personal project — creation of a training hospital specializing in mental illnesses for residents in the District of Columbia, and men in the armed forces. Thus was born St. Elizabeth's in Washington, still regarded as one of the world's best. She designed the hospital and selected the superintendent.

During the next congressional summer recess, she vacationed in Nova Scotia. On an earlier holiday there, she had interested community leaders in the need to construct "Hospitals for Mental Diseases." As usual, men in the provincial legislature had no great interest in the plight of non-voters. Though professional men — doctors, prison and hospital administrators — were almost unanimous in support of Miss Dix's plan, lawmakers' interest was tepid; nothing happened. When she returned to Halifax in July, 1853, she ferried one day to Sable Island, famed for its fiercely picturesque coast. Her grandfather and brother had been seafaring men so she noted in her diary an anomaly: "no Life-Boats, no Fog-bells, no Lighthouses." The very next day, a brand-new boat sank there in a stormy sea. Dorothea canvassed friends in the U.S. and England, and raised a fund to buy six new vessels, specially-equipped for the dangers of the channel. She herself negotiated with the builder a discount on each boat. The Dix Bill, which had languished in the provincial legislature for years, rode through with almost no opposition, and money was appropriated to construct what still stands — and is now known as "Mount Hope," overlooking Halifax Harbor.

It was much more difficult in Washington. Each new Congressional session in Washington faced increasingly complex conflicts with regard to homesteading. Powerful interests, making a grab on public lands, proliferated constantly: railroads wanted to expand their trackage; settlers from the east wanted some; land speculators and industrialists investing in mineral resources; veterans groups demanded 160 acres for each survivor of the war. "Miss Dix's Bill," with its simple humanitarian objective to allocate 12,225,000 acres of public lands — however nonpolitical and moral the case — no longer would find still waters in which to navigate. Nevertheless, the Senate, traditionally more responsive to principles and less to politics than the House, voted overwhelmingly to approve the bill at the start of the 1854 session. It looked like a sure thing. Printing of 5,000 copies of the Bill was authorized to be circulated nationally. The new law only had to survive debate in the House of Representatives.

Stephen Douglas chose this moment to introduce his Bill concerning the future of the Kansas-Nebraska territory. The slave issue again threatened to sabotage action. But Dix had by then built such pressure that her land-grant Bill squeaked through. Victory!

It was not to be. Mississippi's Jefferson Davis, Secretary of War, was powerful in the cabinet. President Franklin Pierce was tiptoeing on eggshells to avoid federal provocation of political leaders in the South. He vetoed the Bill, and Dix narrowly failed to muster a two-thirds majority to override.

The greatest enterprise of her life having been rebuffed in the 11th hour, Dorothea suffered a period of rage, of disgust with the political system; of profound depression. September 2, 1854, she set sail for Liverpool and the happy environment of the Rathbone estate. England was entering a turbulent period of population migration. Miss Dix resumed her traveling and soon came to realize in "progressive" Britain "there are so few ways in which the laboring classes can benefit themselves ... there are no middle classes ... and the rich control the poor beyond any power of the laborer to shake off," she wrote to Fillmore.

Dorothea's travels and investigations of institutions housing unfortunates, eventually would cover the entire British Isles. In desperately poor Ireland, she investigated hospitals, prisons, workhouses and almshouses, averaging one a day, sleeping in a different town every night. Scotland provided a completely different, but equally serious, situation. Like the American South, Scotland jealously cherished autonomy from federal control. When England, decades before, had enacted its historic "poor laws" - which included provision for lunatics — Scottish leaders had rejected them. They preferred the feudal tradition, whereby nobility in each county appointed a sheriff who was given autocratic power. He had sole authority to authorize and license private madhouses, to inspect and report on them when he chose, to appoint a medical supervisor for each, to allocate funds. In the town of Musselburgh, Midlothian county, near Edinburgh, Dorothea saw with her own eyes that the sheriff was a scoundrel. He selected madhouse-operators that were "people of the lowest grade of character, and very ignorant" she wrote. They, in turn, hired the "supervising physician." This individual thereafter was in thrall, happy to report, regularly, that all was well. Supplies were cut back; the cheapest food was served in smallest possible portions. Cost-cutting in the interest of greater profit was the order of the day. Law "required them to report abuses," Dix wrote, but "their pecuniary interests urged them to (overlook) ... the proprietors had things all their own way and they were intent on making money."

As a foreign national, Miss Dix had till then refrained from speaking publicly or writing about her observations in the institutions visited. Flagrant dishonesty and merciless exploitation of people who could not defend themselves were too much for her to bear. Though independent in their authority, sheriffs were subjects of the English monarch as "officers of the Crown." Backed by physicians in Edinburgh to whom she reported the conditions in nearby Musselburgh, she wired London "to demand of the Home Secretary ... a Commission for Investigation."

She traveled to London by overnight train. Lord Shaftesbury had summoned six other members of the Lunacy Commission, doctors and lawyers to the meeting. Dorothea described her personal investigation, and provided details of outrage. The sheriff had poo-pooed her shocking observations; she had been insulted and dismissed. She characterized him as an evil, selfish man, who had "trifled, jested and prevaricated."

Dix was famous worldwide as a nonpolitical, objective authority on the care of citizens incapable of caring for themselves. Normally formidable in her personal bearing, in this appearance before the Lunacy board, she was feeble from illness and fatigued from her all-night journey. The Lunacy Board recommended to the Home Secretary, Lord Grey, that he hear her complaint. He did, and was persuaded to endorse her petition for an investigation. He advised a toning down of her indignant rhetoric, however, in favor of stressing the conflict of interests built into the Scottish system. He guided her through the tangled politics of relations with officials, so that even the Scottish Members in the Parliament were won over during her lobbying efforts.

The Dix resolution passed in record time. Under the authority of the Crown of England, assuming seniority over the Great Seal of Scotland, Royal Commissioners were not only to enquire into "the state of the Lunatic Asylums in Scotland" but beyond that, into the larger question of new laws "respecting Lunatics and Lunatic Asylums in that part of the United Kingdom." Queen Victoria signed the edict. Dorothea Dix chose the members of the Commission.

The American reformer then went to the British islands of Jersey and Guernsey, did her investigative work, proposed reforms, got them approved. She had earned her vacation on the Continent of Europe. David Gollaher says "in the space of a year she would cover perhaps ten thousand miles using ... deluxe steamers, crude barques, rowboats, trains, carriages, wagons, carts pulled by beasts ... even a camel ... she often made her way on foot," visiting madhouses, prisons, hospitals even those in remote areas that required cross-country hiking. In Rome, an audience with Pope Pius IX effected reforms in church institutions and construction of new facilities.

From Italy to Greece, to Turkey, north to Hungary, east to Russia, west to Scandinavia, south into the Low Countries, west again to France, back across the

Channel to England and the Rathbones. Only then did she rest, before return to America.

Were the story to end here, it would record a full and successful life. But there was more to come. The American Civil War.

It should have been the capstone of her career; it was, instead, an anti - climax. An admirer once said "You are a moral autocrat; you speak and your word is law." The "moral autocrat" rushed to Washington at the outbreak of the war. She reported to the Surgeon-General of the Army and volunteered to organize a women's nurse corps. In this ambition, of course, she was emulating her idol, Florence Nightingale, whom she had gone to visit at Scutari, in Turkey. She was named Superintendent of Women Nurses in the Union Army.

Like the British on the eve of the Crimean War, the U.S. Army was ill-prepared. Regular forces numbered 16,000 men; no army hospital contained more than 40 beds. An enormous effort was called for. Dorothea's power was great — to select and assign women nurses to military hospitals. That power made it possible for defects in her character to become obvious. Yes, she was absolutely trustworthy in the handling of large sums of money that were contributed by well-wishers. But no, she would not accept into her corps qualified women who she did not think were "sober, earnest, self-sacrificing (able to) exercise entire self-control." Her edict: "No woman under thirty need apply ... (they must be "plain looking.") Her idea of a "qualified" woman was a mirror-image of herself. Perhaps that description was necessary for a female to survive in army hospitals — doctors and government officials were disrespectful, intolerant, abusive and chauvinistic. Without doubt, however, many women who were well-qualified by experience and temperament to be of great help in the awful battles that followed during four long years, were barred by Dix from serving despite the desperate shortage at all times.

A few of great determination and integrity bypassed the Dix barrier: Clara Barton, founder of the American Red Cross; Mary Ann Bickerdyke ("Mother" Bickerdyke who closely followed Grant's and then Sherman's campaigns) and many others were too independent to take orders. They did heroic service in field hospitals while Miss Dix, in her sixties, confined herself mostly to exercising her authority in and near Washington; the farthest afield she went was Fort Monroe on the Potomac.

She resigned her commission as of September 1, 1866, after Lee's surrender at Appomatox Courthouse. She raised a fund from friends to erect a 45-foot obelisk at the Fort Monroe cemetery where, she wrote, "are interred the remains of 6,000 American soldiers who laid down their lives ... to maintain the laws of their country." Secretary of War Stanton sent a color guard to solemnize the dedication, attesting to her "arduous, patriotic, humane, and benevolent labors."

After the war, she devoted herself to tidying up its detritus: helping orphans and widows, damaged veterans, missing men and women — including the redoubtable Mary Bickerdyke, who had been, at war's end, a one-woman nursing corps for starved, near-dead survivors of the prisoners-of-war hell-hole at Andersonville. After which, there was no place for civilian Bickerdyke in America's medical system. She could not find work as a nurse; vanished for five years, was found — by then demented – a charwoman in Bethany Hospital for the Insane in New York. She was rehabilitated by Miss Dix.

Dorothea remained busy, continued to agitate for mental-institution reforms, when called upon. She intervened often in behalf of the two-dozen asylums she had founded, that were being starved for funds by state legislatures. Despite her well-known services to the Union Army, politicos even in the South held her in high esteem; her portraits were commissioned to hang in many public institutions there, as well as the north. She continued to travel, including a rail trip to California, when the "golden spike" connected train service all the way.

Loans to her brother's family over the years, never repaid, totaled $50,000, greatly depleting her capital. She further invaded her capital to send gifts to the institutions she had founded. She was generous to animal causes. She paid for a watering-fountain for horses built in Boston, for which the poet John Greenleaf Whittier provided the text for the engraving at its base:

> Stranger and traveler
> Drink freely, and bestow
> A kindly thought on her
> Who bade this fountain flow,
> Yet hath for it no claim
> Save as the minister
> Of blessing in God's name.

Dorothea Dix died July 17, 1887 after six years of residence at the New Jersey State Hospital near Morristown, which she had been instrumental in founding.

CHARLES DREW
Pioneered the Struggle to
Emancipate Black Physicians

Speaking on the floor of the House of Representatives, May 28, 1942, Congressman John Rankin of Mississippi opined that

> "one of the most vicious movements that has yet been instituted by the crackpots, the Communists, and parlor pinks of this country is that of trying to browbeat the American Red Cross into taking the labels off the blood bank they are building up for our wounded boys in the service so that it will not show it is Negro blood or white blood. That seems to be one of the schemes of these fellow travelers to try to mongrelize this Nation ... Thank God, the Red Cross has stood its ground and refused to permit this outfit to have Negro blood pumped into the veins of our wounded white men on the various fronts ... They seem to have some crackpot alien doctors advising them that it makes no difference what race this blood comes from."

The most prominent of the "crackpot alien doctors" who had asserted there was no scientific difference between the blood of a white man and one of color was Charles R. Drew, founder of the American Red Cross Blood Bank program and its first director. He was a Black. Though John Rankin's fulminations on the floor of Congress clearly were an extreme of bigotry and stupidity – and were recognized as such – the ingrained race prejudice it expressed was common in America at that time, particularly among elective officials in southern states. Mississippi's Senator Theodore Bilbo's book "Take Your Choice: Separation or Mongrelization" thundered "Nothing is more sacred than racial integrity ... when the blood of races mix, the white blood ... is forever lost." The American Red Cross was simply implementing the adamant race-separation instructions of the surgeons general of the Army and the Navy. A similar prejudicial attitude prevailed in the groves of academe, in finance and the corporate world, in government, the professions, in trade unions, and in the councils of America's professional associations. The barriers to black equality in any field of endeavor were high and solid. Most of his life, Charles Drew devoted to scaling those walls in his field, medicine. His last years were devoted to leveling them; but his most substantial successes were not realized until after his death.

His father was "blue collar." His mother was that rarity – a female, black, college graduate. They lived in a racially mixed, ethnically diverse neighborhood, in a rigidly segregated southern city, Washington, DC. By decision of the Supreme Court in 1883, racial discrimination was perfectly legal. The Washington "Bee," a black-owned newspaper, termed the city a "citadel of race prejudice ... (ruled by a) white American civilization (that) was relentless and often bitter." Charles grew up in that environment; the hardships of his life were likewise relentless and often bitter.

The Drew parents and all the Drew children were light-skinned, descendants from mixed-blood unions generations back. They could have "passed" in white society, but never wished to. (In fact, Charles made it a point to identify himself as a Black in social and professional concourse.) He attended public schools and participated in athletics from an early age. At eight, he was a prize-winner at city-wide swim meets. He worked at summer jobs and delivered newspapers, having organized a team of 10 friends to cover all nearby neighborhood routes. At his high school, he earned four varsity letters and medals for excellence in athletics in his junior and senior years. Looking back, Drew's brother comments "He was always very ambitious and restless ... He had a tremendous will to win, to get ahead. It devastated him if he didn't win ..." In Drew's yearbook, under his photograph, he is quoted: "You can do anything you think you can."

From 1922 to 1926, he attended Amherst College, then as now a leading liberal-arts institution devoted to social progress. As before, he was active in sports – so active that his grades suffered and he was admonished by the dean, "Mr. Drew, Negro athletes are a dime a dozen." Which didn't prevent the lad from winning letters in track as early as his freshman year. He was nominated for the All-American football team in his junior year. His college coach recalled

> "In all of Amherst College's long history, no campus generation treasures a more glorious football memory than the graduates of 1923-26 ... (the) most memorable thrill while I coached there was given by a tall well-built Negro halfback from Washington, D.C. ... Charles Drew."

But – as one of only 13 blacks in a student body of 300 in a school concerned with social progress – his starring record didn't weigh sufficiently to be named captain of the football team in his senior year.

Charles' first job out of college, in 1926, was at all-black Morgan State College in Baltimore, MD, teaching chemistry and biology; and serving also as Director of Athletics. Which was a challenge; the Morgan State teams, when he arrived there, were often defeated by high school opponents. Charles trained them into becoming winners.

But coaching was not in his mind as a career goal.

Originally intending to become an engineer, as his perspectives widened, he became aware of the color-barrier in that profession. There weren't many openings for black engineers. He switched track. Negro doctors were honored in black communities and often achieved affluence. He applied to Howard University Medical School, one of only two in the nation that accepted black students. But the Admissions dean was dissatisfied with insufficient and inferior Amherst credits in liberal arts. (Says W. Montague Cobb, at Amherst with Drew, he "never got a mark better than C in college.") He applied and was accepted at McGill University Medical School in Montreal, Canada, which was known to be more friendly to blacks than their counterparts in the United States.

At McGill, he became captain of the track team in his second year and won Canadian championships in the high and low hurdles, high jump, and broad jump. By this time, the single-minded focus upon athletics that had prevailed at Amherst was a mistake he didn't intend to duplicate. Drew now concentrated on his grades, too. He won awards in academic and medical specialties, was named to the honor society, worked on the school newspaper. By graduation in 1933, he was second in a class of 137.

But he nearly didn't make it through his senior year. Color was no problem. Money was. The Great Depression was on, and his father, the family breadwinner, had been laid off from work. The despondent student faced a dilemma. He wrote his family

> "Here I am, a stranger among strangers in a strange land, broke, busted, almost disgusted, doing my family no good, myself little that is demonstrable. Yet I know I must go on somehow – I must finish what I have started ..."

In the nick of time, he was awarded a $1,000 fellowship from the Rosenwald Foundation. It funded tuition and books the last year, relieving the pressure on family finances. After graduation, he wrote to the Foundation: "It is my constant hope that I shall be able at some time to add some new thought, discover some new process, or create something which will prevent or cure disease, alleviate suffering or give men a chance to live ... And thereby in part repay the debt which I am happy to acknowledge."

Upon graduation from McGill, Drew entered Montreal General Hospital as an intern; and then continued there as a resident physician in internal medicine. His academic record was excellent; he was one of the three top students in his group. During this intensive period, he had his first experiences with cross-matching blood types prior to giving transfusions and became interested in hematology, as vital to the practice of surgery. He earned a master's degree.

Surgery was in his mind when the time came to choose the medical specialty in which he wished to carve a career. It would have been a natural progression go on for a Ph.D. He was refused admittance at several medical schools to which he applied. It was a bewildering rebuff, considering his superior performance at

McGill and at the Montreal General. He could not understand the unanimous rejections. It couldn't be simple color prejudice – not at such high levels in the medical Establishment.

Or could it? He learned, years later, that it was color, all right – but of an unusual shading. Drew had by then been accepted into the surgery department at Columbia on a fellowship awarded by an outside philanthropy. The bizarre reason Drew had initially been rejected when he applied to Columbia emerged in a casual conversation with Dr. Allen O. Whipple, chairman of surgery, who had blackballed his original application:

> "I am certain you could do well with the average patient, but could you, with your background, feel at ease and render competent service if one of your patients in surgery were a Morgan, an Astor, a Vanderbilt ... We have such patients here. In the selection of our residents, we choose only from among superior students and we take into account their family and personal background."

It was a crushing realization. Because he was black, Drew could hope for further training only in a black medical school. But there were none at the level he wanted.

The elder Drew had died in the depth of the depression; Charles, as the oldest among his siblings, would have to assume head-of-the-family status, and contribute to family income. So it was that, instead of continuing with his Ph.D. education and certification as a specialist in surgery, he had to be content with the master's degree earned in Montreal, and accept an entry-level position as an instructor in pathology at all-black Howard University Medical School – where his color was not a barrier, but a passport.

He taught at Howard for three years, but never shelved his ambition for specialist study. As often happens with a well-prepared mind, a lucky break came his way. The Rockefeller Foundation, a philanthropy that pioneered in providing advanced training opportunities for minority candidates in science and medicine, awarded him a fellowship for continuing study at Presbyterian Hospital, affiliated with Columbia University in New York – under Dr. Whipple! Said a colleague at the hospital: "He persuaded Whipple to train him as a resident ... a bootleg residency ... Surgical residencies for blacks did not open up till the late 1940s ..." He worked at Columbia for two years, from 1938, stretching the family budget into penny-pinching allocations for rent, food, transportation; the balance from his stipend went home. It was a grim life of grinding poverty. In the spring of 1940, he commenced work on his Ph.D. dissertation, "BANKED BLOOD: A Study in Blood Preservation." It embraced all he had learned to that point in the laboratory and in clinical practice; bulked to hundreds of pages.

Once again, Lady Luck intervened in the form of a cablegram. He had studied at McGill with a young English bacteriologist, John Beattie; and later worked under him as an intern and then a resident physician at the Montreal General Hospital. They had collaborated with research into stabilizing and preserving blood for transfusions. Beattie was now back in England, vetted as lieutenant colonel in the army. England was preparing for war with Germany; there would be need for huge supplies of blood for transfusions – far beyond what could be expected from the whole-blood supply system. Beattie was on the British planning group preparing for that day, canvassing contacts in America for assistance. His wire to Drew asked for 5,000 ampoules of dried blood plasma. It was the knock of opportunity for the young man. With the participation of his superior in the lab at Presbyterian, John Scudder, they set up a Blood for Britain collection facility at their hospital, August 15, 1940. It quickly spread to eight other hospitals in the New York area. In little over a month, 1,300 donors were being processed weekly. According to Red Cross historian Clyde Buckingham,

"Six weeks after he began work as medical supervisor for the Plasma for Britain project, Dr. Drew had become an acknowledged leader in the program."

Drew and Scudder followed through by writing a proposal to their hospital superiors for funding in order to establish a permanent blood bank. It was approved. Despite their concentration on blood-science, at the time, Scudder and Drew were very junior among a number of physicians all over the world who were seeking to systematize procedures and standardize specifications for stockpiling blood. The pair at Columbia soon directed their research into extracting from whole blood just that portion which could be preserved. This plasma is the ingredient in whole blood which contains red corpuscles. It can be administered without concern for the matching of blood types. While whole blood has a brief shelf life, there was good reason to believe that plasma could be stored for long periods, possibly indefinitely. Within six months, Scudder and Drew forged to the forefront as worldwide authorities on blood preservation.

During that feverish period, Fate had again stepped in. On a brief travel-stop at Spelman College in Atlanta, he had met and fallen deeply in love with beautiful Lenore Robbins, a home-economics instructor on faculty there. He conducted a whirlwind courtship and, within five months, the couple was married, September 23, 1939. They set up housekeeping in a small apartment near Columbia – shared with another couple. Lenore got a part-time job at Columbia. But planning for the future required a larger, stable income. So, at the conclusion of his two-year fellowship in June, 1940, Drew accepted reappointment to the Howard University Medical School with a modest promotion to assistant professor of surgery.

In September, 1940, the war in Europe was going badly against British forces and creating urgent need for blood plasma in huge quantities – far greater than existing sources could supply. Drew was summoned back to New York immediately. A leave-of-absence was granted at Howard University.

From September, 1940, Drew headed the Blood for Britain project in the New York area, major collection center for blood deposits in the nation. He devised the logistics to collect whole blood, extract the liquid plasma, and ship it, refrigerated, to Europe, where the recipients would test consignments to be certain they were sterile and stable despite the long process of handling, storing, and the agitation of the journey. The work he and Scudder had pursued, only months before, was chosen by the American Red Cross as holding the greatest promise for future operation of blood banks.

Starting in 1938, at Presbyterian Hospital before the war, Scudder and Drew had pioneered in studying and documenting fluid loss and blood volume in treating critically ill patients. It was an efficient transition to preparing technical specifications for extraction of liquid plasma from whole blood on a high-volume, round-the-clock basis—and then to transform it into dry form. "Thus was set in motion the national project for the production of the No. 1 life-saving agent of the war ... dried plasma," testified William DeKleine, who was medical director of the American Red Cross in 1940 and 1941.

Starting out as equal partners, Scudder and Drew wrote a series of technical papers that became a "Bible" for blood-collection centers internationally. Drew gradually became team-leader; the last group of papers issued from their office carried only his name as author. Scudder has written that his partner was

> "naturally great. A keen intelligence coupled with a retentive memory in a disciplined body, governed by a biological clock of untold energy; a personality altogether charming, flavored by mirth and wit, stamped him as my most brilliant pupil. His flare for organization with his attention to detail; a physician who insisted upon adequate controls in his experiments ... (He became) one of the great clinical scientists of the first half of the twentieth century."

The surgeons-general of the U.S. Army and Navy signed on. Guided by the Drew-Scudder directives, they launched programs to create support facilities for emergency blood plasma transfusions on and near the fields of battle. The pair formulated procedures so that plasma could be kept stable for at least two months of handling and shipment. Soon regular consignments were going to the British, though the U.S. was not officially at war with the Axis powers. The receiving officials in England reported that shipments of liquid plasma, and then dry plasma, were arriving in perfect condition, ready to use.

Within weeks, Drew was appointed the first Director of the American Red Cross in New York City, which became the keystone of the program that thereafter was responsible for collecting blood for American armed forces. In its comprehensive report on World War II blood programs, the U.S. army, through its historian, Brigadier General Douglas B. Kendrick, testified that "The experience of the New York Chapter served as a pattern for the organization and operation of the (entire) Blood Donor Service."

Anticipating the day when the U.S. would join the war, Drew commenced circularizing hospitals and commercial laboratories in 12 major U.S. cities. He recruited allies willing to collaborate in programs for production, storage, and shipment of dried plasma. In December, Drew was able to submit a detailed manual of operation, and requisition for funds to organize a donor program that would supply Britain's needs while simultaneously stockpiling reserves in anticipation of America's inevitable entry into the war. The vast undertaking was implemented February, 1941, in New York. Drew even supervised the modification of trucks to become mobile donor stations. Within 60 days, his program was scheduled to go national. Official archives of the American Red Cross acknowledge that Drew was the nation's foremost

> "expert on methods and techniques of blood procurement and processing. He ... brought together, for the benefit of hematologists everywhere, the latest knowledge acquired by scientists ... the results of research studies and clinical tests of academic and commercial laboratories on both sides of the Atlantic ... a major step forward toward the goal of mass production of blood supplies in the large quantities and in the form which could be used by the armed forces under conditions of modern warfare."

In a six-month period, Drew had become internationally famous, hailed as a brilliant scientist, an inspiring leader and a highly efficient chief executive. He wrote to his wife: "I have made contacts that I may have waited a lifetime to make in the ordinary scheme of things ... "

What type of contacts? Did he have a specific job in mind? Close friends believe that he was slated for national office within the American Red Cross. He was certainly the most qualified candidate for the post of national medical director and chief operating officer of the blood program. It is believed that he expected such an appointment. But 1941 was premature in the history of the Civil Rights movement. There were almost no blacks in top-echelon jobs with private industry, in government, the armed forces and even in service

79

organizations – other than those in the Negro community. He was not offered the national position that he was, uniquely, the best qualified man in the country to occupy.

Despite pride in the obvious success of his program, and although full implementation still had some distance to go, Drew felt obliged to resign his post as medical director. As he wrote to his mother, finances were in a "morass." His wife had moved back to her family home. He couldn't afford further time in public service, lacking substantial salary support and career security. He left the job a month before his Howard University "leave" was to expire.

On April 1, 1941, he appeared before the "oral examiners" of the American Board of Surgery, and was duly certified as a specialist diplomat. His "exam" turned into a lecture that held his judges transfixed far longer than the time allotted. (Within months, he was invited to become an "examiner" for future candidates!)

He then crammed for his "orals" for the Ph.D. at Columbia, updating the work commenced earlier at Presbyterian Hospital for what became his "Banked Blood" treatise - embracing up-to-the-minute data prepared for papers he had worked on during the preceding year. Granting of the degree was almost a formality.

Notwithstanding the eminent "contacts" he had made at national and international levels, the Drew family moved back to Howard University in October, 1941. Charles was appointed Chairman of the Department of Surgery and Chief Surgeon at Washington's all-black Freedman's Hospital. A few weeks later, the American Red Cross announced a new national policy – to exclude black donors form the blood-bank program; the very program that a black physician had been the key figure in setting up. Even blacks in the armed forces were turned away from donor stations, as were black civilian applicants. Did Charles Drew have an inkling that this policy decision was "in the works" while he was still working for the Red Cross? Had it been a factor in his resignation?

The strong anti-black sentiment for which John Rankin was a spokesman cheered. Exclusion of "black blood" from the humanitarian program was Senator Bilbo's assurance of racial purity. A storm of public disapproval followed.

When the U.S. entered the war – and black soldiers were as much at risk as white – the Red Cross reviewed its controversial blood-donor policy in conference with the surgeons-general of the Army and the Navy. On January 21, the Red Cross announced that it would accept blood from black donors – but would segregate it. The press howled! Typical were the comments of columnist Albert Deutsch, writing in the New York PM Daily, March 30, 1944:

> "At the very time Drew was setting up the Red Cross Blood
> Bank, helping to save thousands of American lives through his
> brilliant scientific ... work, his blood would have been rejected

by the American Red Cross ... Later, when the Red Cross modified its policy and accepted Negro blood ... his blood would have been segregated..."

Charles Drew, recognized as the nation's leading authority on the blood program, voiced protest:

"the recent ruling of the United States Army and Navy ... is an indefensible one from any point of view ... there is no scientific basis for the separation of the bloods of different races."

The black press was vociferous in its condemnation of the armed forces' discrimination in the blood-donor policy – and critical of Dr. Drew's low-key reproof. In the months and years that followed, during the course of World War II, Drew tried to influence the War Department's policy by arguing logic and reason. Reason didn't work any better than logic or the facts of science. His position stiffened and finally, in 1944, he fired a tough statement:

"I think that the Army made a grievous mistake, a stupid error in first issuing an order to the effect that blood for the Army should not be received from Negroes. It was a bad mistake for three reasons: (1) No official department of the Federal Government should willfully humiliate its citizens; (2) There is no scientific basis for the order; (3) They need the blood. I would be heartily in favor of pressure of all types being brought on the Surgeon General of the U.S. Army to force him to rescind the instructions to the American Red Cross, which demands the separation of the blood of donors. Such an order, I believe, would have a greater salutary effect upon the morale of Negroes than any other act which could be done at this time. I have had occasion to say this before both official and unofficial in Army circles and outside of it, but to date, with no avail."

Thousands of protest letters were received by the Red Cross, the President, and others. Typical was that from Walter White, executive secretary of the National Association for the Advancement of Colored People: "How ironic must be the laughter ... in Berlin and Tokyo as they listen to American assertions that the war is being fought against racial ideology of Aryanism and to wipe out ... racial bigotry!"

Emergence of Charles Drew as public spokesman for the anti-segregation movement, rather than a well-behaved, mild-mannered organization man

81

probably dated from this time. He had every reason to be angry and discouraged. But he wasted no time lamenting reverses, disappointments, injustices. After all, the phobia opposed to mixing blood was not confined to extremists like John Rankin. No less a humanitarian than the "Great Liberator," Abraham Lincoln, was on record against mixing blood. White physicians were quoted in "Outcasts from Evolution" as recommending castration in order to deal with the animal passions of the Negro and render them "docile, quiet, and inoffensive." This was proposed in the "Atlanta Journal Record of Medicine" in 1906. It was a widely-held attitude, not a crackpot outburst.

Noteworthy was the award to C. Canby Robinson in 1948, the (white) director of the Red Cross blood program, of the organization's Medal of Merit. No such recognition ever went to the program's founder, Charles Drew. The official history of the blood program, published by the U.S. Army in an 810-page book, never mentions Drew once. Dr. William DeKleine, from 1928 to 1941 Medical Director of the American Red Cross chimed in: "Dr. Drew had no relationship whatever to the blood plasma project for the military forces, which was started shortly after the London project was discontinued." When DeKleine died in 1957, the "New York Times" obituary awarded him sole credit for the Red Cross blood bank program during the war.

Drew recognized that making progress in causes of importance would come only by focusing on the things he could change. To his high school coach, he described a new personal crusade he had undertaken:

> "at Howard in the Department of Surgery ... the situation is comparable to the sport situation when I took over at Morgan State College ... Seventy years there has been a Howard Med School, but still there is no tradition; no able surgeon has ever been trained there ... In American surgery, there are no Negro representatives; in so far as the men who count know. All Negro doctors are just country practitioners, capable of sitting with the poor and the sick of their race, but not given to too much intellectual activity and not particularly interested in advancing medicine. This attitude I should like to change."

The crusade that Drew undertook was simply to upgrade the standards under which black medical students trained; and to oppose, vehemently, the bias, disrespect and discrimination with which black medics were treated within the profession.

He wrote to the editor of the "Journal of the American Medical Association," charging that institution with

"One hundred years of racial bigotry and fatuous pretense; one hundred years of gross disinterest in a large section of the American people whose medical voice it purports to be ..."

When the A.M.A. apologized, pointing out that membership in the association could only come through belonging to a county medical society, Drew counter-punched:

"You know and I know that it is utterly impossible for a Negro physician to become a member of a county medical society in the South ... It is a cause of repeated humiliation. It is a constant indictment of the principles on which the American Medical Association is supposedly founded."

Drew was determined to transform Howard Medical School – and black medicine nationwide – into a model for future generations. He canvassed white medical schools and white hospitals all over the country, keeping up the pressure to provide internships and residencies for black medical school graduates. He challenged the white hospitals of Washington, which had long excluded black physicians from seeing their black patients – and the few that would admit black patients, but did not permit their doctors to treat them.

He denounced restrictions barring admission of black physicians into national and local professional associations. He was an agitator, a vigilant dissident, shaking the foundations of status quo. With solid statistics attesting to the superior quality of the medical students he had trained, he slowly made progress in breaching the ancient walls of anti-black prejudice. In Washington, where his influence was powerful, of the 200 licensed black physicians, 26 became board-certified specialists, one-third of the national total for blacks.

He threw himself into battle with the U.S. Congress and the medical establishment of the D.C., demanding rehabilitation and remodeling of Washington's Freedman's Hospital, which served the black community. It had been antiquated for years, and was desperately short of modern equipment. He fought for expansion, as the hospital was perennially overcrowded.

He traveled widely and often, raised funds, shaking the tambourine for money and fellowships for black medical students. He had no hesitation pulling strings and using influence with contacts all over the country, even abroad, to get help and equipment when it was needed. He continuously upgraded training standards for Howard students so that, when they were accepted as interns in white hospitals, they would be recognized as outstanding. He came into conflict with the Howard University faculty power structure for sending elsewhere the medical school's most promising students. Because his objective was broader, he took the flap in good spirits and continued, undeterred, on his chosen course.

His rewards were the evidence of improving acceptance statistics for black candidates at white medical schools and white hospitals across the country. "Drew-trained men were among the earliest (black) pioneers in gynecology, neuro-surgery, cardiovascular surgery, and other fields ...fashioning new images for blacks in surgery" reports Dr. Charles Watts, a former student. When Drew sent a group from his first class of graduate students to Johns Hopkins for exams to qualify for certification by the American Board of Surgery, in competition with top students from top white medical schools in the country, they were graded first and second among their peers.

Little interested in the appeal of social prominence and affluence that usually rewarded a black physician in a black community – certainly one as nationally famed as he – Drew chose the path of militant anti-materialism. He never owned a home, contented with the old, three-story shambling structure provided for him on the Howard campus. He accepted no private patients, no lucrative consultancies, satisfied with the modest salary the college and hospital could pay, on which he and Lenore reared a family of four children. He worked a 14-hour day normally, every day but Sunday, when he took off in the afternoon for family relaxation and gardening. His wife recalls: "Even when he would sit down with the children, he wanted them in possession of something when he got through with them. He had many other financial offers – he could have had his own lab and done research – but he wanted to train black surgeons. He felt he could do the most good that way."

His standards were high, souvenirs of his training at Amherst, McGill, and Columbia. He insisted that Howard medical students be well-groomed, dress well, do well in their academic subjects as well as scientific. He kept in touch with many of the young men when they left Howard, advised them on career choices, even mentored their personal problems. At all times, he was mindful of the missions he had undertaken, a calling greater than the pursuit of his personal career. He wrote to Jack White, a former student:

> "Our horizons are being widened by the residents all the time and the things they write back, sharing with us ... their daily experiences ... enriches us all and at the same time forges the bonds which unite us even more firmly so that each man is inspired to do more and more on his own in order to be worthy of the fine companionship of such a group. In the individual accomplishments of each man lies the success or failure of the group as a whole. The success of the group as a whole is the basis for any tradition which we may create. In such tradition lies the sense of discipleship and the inspiration which serves as a guide for those who come after, so that each man's job is not just his job alone, but a part of a greater job whose horizons we,

at present, can only dimly imagine for they are beyond our view."

In collecting evidence like these quotations, Spencie Love perceptively illuminates the premise inherent in the subtitle of her biography: "One Blood: The Death and Resurrection of Charles R. Drew." (The subtitle is, in fact, the apotheosis and significance of Drew's life.) The practical measures he undertook in the last years of his life at Howard Medical School, and his beavering away at the wall of prejudice against black physicians within the medical establishment started to show results beyond his horizons. They have expanded significantly in the years after his premature death.

In his mid-forties, Drew was in excellent heath, years of competitive athletics having built a heavily-muscled frame. His discipline of long hours of work and study ruled his body; he slept but a few hours a night. Hence his decision, undertaking a long auto trip to a medical convention in Alabama, to drive all night on the first leg the of route and thus solve the problem of scarce motel accommodations that would accept blacks in the Carolinas. He didn't start out till after midnight, Friday, March 30, 1950, having been the featured speaker at a student banquet. After checking patients at the hospital, he picked up passengers – a doctor colleague and two interns. His wife had urged him to fly the next morning, but Drew was mindful that the interns could not afford the airfare, and he was anxious that they attend the convention.

At the wheel of the car, near the rural community of Green Level, North Carolina, Drew apparently dozed off as the dawn was breaking on the first day of April. Veering off the road, Drew attempted to correct the course. Too abruptly. The violent swerve caused the vehicle to overturn. He was half-thrown out and the heavy machine rolled over him. Neighbors called the highway patrol, and ambulances arrived within minutes. Less than an hour after the accident, the four men were being treated at the Alamance County Hospital. Three were conscious and identified themselves. Only Drew had been so badly injured as to be unconscious. The hospital emergency room worked feverishly, but his wounds were mortal. He died on the operating table.

Robert Jason, Dean of the Howard University Medical School, eulogized "Perhaps it was because of the brilliant glare of ... vision in his mind's eye that he taxed his physical endurance too much and he closed his eyes."

The Alamance Hospital, the only one in the rural county, was for whites, but every attention possible was lavished upon the accident victims. Despite which, the black press of North Carolina spread a story that Drew had died because he was denied blood in a segregated hospital. There wasn't a word of truth to the story. The three survivors from Howard Medical School filed refutations of the story. Nevertheless the fiction quickly spread into a legend which still lives. Doing research for this essay, the writer encountered it on the Internet. A half-century after his death, the false legend that Drew died "because he couldn't get blood" is still widely believed in the black community, says that American Red Cross, and greatly affects blacks' willingness to donate blood – even for causes known to be Afro - American.

The final word belongs to the eminent French seer, Fernand Braudel, in "A History of Civilization." He remarks that man's " history of events ... social history, the history of groups and groupings" are marked "by slow, but perceptible rhythms ... in his relationship to his environment." The rhythm set in motion by Charles Drew gathers strength with time, after his death. Schools and bridges are named Drew. Scholarships are awarded from funds honoring him. Los Angeles has the Charles R. Drew Postgraduate Medical Center, affiliated with the University of Southern California and the UCLA medical schools. Portraits of Drew hang in the National Institute of Health in Bethesda, MD, and the American Red Cross headquarters in Washington, D.C. two years after Drew's death, black doctors were admitted to the A.M.A. chapter in Washington.

The twelfth man
blew the whistle
RICHARD FEYNMAN

On January 28, 1986, a huge space shuttle roared aloft from its launch pad at Cape Canaveral, Florida, and – barely a minute later – exploded. The passenger capsule, intact, plunged into the sea, killing the captain, Francis Scobee; schoolteacher Christa McAuliffe; pilot Michael Smith; technical specialists Ellison Onizuke, Judith Resnick, Ronald McNair; and aeronautic engineer, Gregory Jarvis.

The mission had been postponed four times before that day – and several times that morning – to repair technical problems. Further delay was out of the question; nobody at NASA had the stomach to thwart the President of the United States. It was known that he wished to highlight his "State of the Union" speech that very night, before a Joint Session of Congress, by chatting with the "Challenger" astronauts in outer space.

Hundreds of millions of television viewers saw this most spectacular fireworks exhibition in human history. The tragic event was re-telecast over and over. On following days, additional footage focused on the moment when the fuel tank burst open and the liquid hydrogen exploded into an orange ball of flame.

Why did it happen? So frightful a calamity demanded explanation. President Reagan immediately appointed a commission to investigate. The group, he stressed, was composed "of outside experts, distinguished Americans who have no axe to grind."

Distinguished they may have been, but all those invited to join the commission were federal government employees or pensioners except one – Nobel Prize laureate Richard Feynman. A professor at the California Institute of Technology (Caltech), he had been the youngest member of the "Manhattan Project" (atom bomb) team at Los Alamos. J. Robert Oppenheimer tagged him as "our most brilliant young physicist ... everyone knows this ... a man of the greatest ... responsibility and warmth, a brilliant and lucid teacher, one of the most responsible men I have ever met ..."

Freeman Dyson, also a Nobel Laureate, described him as "having one of the world's most original minds"; Hans Bethe, another Los Alamos teammate and also a Nobel honoree, said "he wasn't an ordinary genius, but a "magician."

In the 40 years since Los Alamos, Feynman broke new ground in physics; revised the syllabus for teaching post-atomic physics; created what scientists today know as Feynman Diagrams. He collected myriad honors and awards;

wrote or co-wrote over 100 papers, many of them creating new horizons in physics. He was awarded the Einstein Prize in 1954 and the Nobel Prize in 1965.

He was famous, also, according to his FBI dossier, as a "screwball." He loved jazz and bongo drums; his hobby was picking locks. He never wore a necktie.

But one piece of information was missing from the dossier. The FBI failed to note that Feynman despised, absolutely hated, the very smell of falsehood, of fakery. As a small child, he'd been taught by his father, a salesman for military uniforms, not to be impressed by outward appearances – what mattered was inside the skin. Richard had quit religious school when he learned that the Bible's morality tales mixed fact and fiction. Having read John Stuart Mill's celebrated essay "On Liberty"; as a college student, he wrote a paper broadening the definition of despotism to include the tyranny of social forms, polite lies, manners in lieu of morals. In English class, he pronounced words exactly the way they were spelled; anything else was an affectation.

James Gleick explains in "Genius," his biography of Feyman that – though he was at the time only a graduate student – he was nominated for the Manhattan Project Team by his department chairman at Princeton because of "unwillingness to accept any assertion of authority ... If there was any baloney or self-deception ... Feynman would find it." After Los Alamos, Murray Gell-Mann, himself a Nobel Laureate, depicted Feynman as "the one physicist ... utterly genuine, free of phoniness, the one who did not worship formalism and superficialities."

So fierce was Richard Feynman's independence that he refused to apply for, or accept, government grants to support his research work.

At 67, battling terminal cancer, Richard Feynman was not flattered by nomination to the "Challenger" inquiry panel and was inclined to refuse. In his mind, politics and science were separate worlds: one was hogwash; the other, holy. He never had endorsed the idea of space travel. ("I never saw in any scientific journal any results of anything that had ever come out of the experiments on the shuttle that were supposed to be so important.") But his wife, Gweneth, argued "if you don't do it. there will be twelve people, all in a group, going around from place to place together. But if you join the commission, there will be eleven people – all in a group, going around from place to place together – while the twelfth one runs around all over the place, checking all kinds of unusual things ... There isn't anyone else who can do that like you can." Still reluctant, Feynman phoned Albert Hibbs, a scientist-friend whose judgment he particularly respected. Hibbs asked "Do you think you could find anything about what happened?" Feynman's answer was "Probably." Hibbs then asked "Do you think it would be important that that problem be solved?" The answer, of course, was "Yes, it would be important." After a moment of silence, Hibbs delivered the crusher: "Why are you asking me?"

After 30 years as a professor at Caltech, Feynman had many colleagues and former students at the nearby Jet Propulsion Laboratory in Pasadena, people who were on the cutting edge of rocket technology. February 4, the day after he became a Commission member, he huddled there with physicists and engineers who had worked on the design of the shuttle, its solid rocket boosters, its engines, and its hardware. He spoke their language, having attended, before Princeton, the Massachusetts Institute of Technology, America's pre-eminent college for training "hands-on" engineers. Summer vacations, Feynman had worked at the Frankford Arsenal in Philadelphia, as a technician.

At the Jet Propulsion Lab, he studied detailed drawings and specifications of every nut and bolt in the Challenger; and hundreds of magnified photographs that had been taken during the launch.

The huge solid-fuel rocket – tall as a 10-story building – was assembled from a series of interconnecting metal sections, each 37 feet in circumference. Joints between them were cushioned by thin rubber rings, somewhat like "washers," 10 feet in diameter. Under the ferocious heat of blast-off, these were expected to expand along with the metal tube sections, and thus retain tight "seals." They were then expected to contract along with the metal when the shuttle entered frigid outer space.

Some of the pictures showed scorching at the joints of the metal sections. Scorching? Feynman noted on his pad: "O-rings show scorching ... Once a small hole burn(s) through generates a large hole very fast! Few seconds catastrophic failure."

He went on: "Zn Cr04 makes bubbles, which means that the zinc chromate putty, packed as an insulator behind the O-rings, makes bubbles which can become enlarged by leaking hot gas, thus eroding the O-rings." Feynman later noted: "The engineers told me how much the pressures change inside the solid rocket boosters during flight ... I learned about the thrusts and forces in the engines, which are the most powerful engines for their weight ever built."

There were other problems with the "Challenger," he learned, such as cracks in the turbine blades of the engines during high-speed tests – failures of hardware. He recalled that when a journalist asked astronaut John Glenn (then a Marine colonel, later a U.S. Senator) what was his last thought before blast-off for the first manned space-orbit of the earth, his reply was "... everything on my craft was supplied by the low bidder!"

Possible causes of the "Challenger" disaster were coming into focus. Feynman decided it would take but a few days to get at the truth, so he carried only an overnight bag to Washington for the Commission sessions.

Prologue to the "investigation" was delivered by the Commission's chairman, William P. Rogers (a former Attorney-General and Secretary of State). They were not "to conduct this investigation in a manner which would be unfairly critical of NASA, because ... NASA has done an excellent job ..."

Another commission member, astronaut Neil Armstrong (the first man to walk on the moon), expressed astonishment that an investigation was thought necessary at all.

Feynman listened – but was of different mind when the hearings began. His credo: "the only way I know to get technical information (is) you ask a lot of questions ... and soon you begin to understand the circumstances and learn what to ask to get the next piece of information you need." He quickly became a "pain in the ass," as Rogers was quoted. When NASA's senior space-flight official, Jesse Moore, was reading his chronologically-organized opening statement, Feynman broke in to note that the temperature had sunk below freezing the night before the launch, and ice had formed on metal surfaces. What effect would very cold weather have on the equipment?

Moore denied that cold could cause problems. He was wrong, of course – and was almost immediately contradicted by another NASA official, Judson A. Lovingood. He testified that NASA officials and engineers at the Morton Thiokol Company, builder of the solid rockets, had conferred at length by telephone throughout the night before the blast-off – increasingly worried about the effect on the rubber O-rings of the plummeting temperature. The Thiokol engineers flatly advised postponing the launch. They were overruled by a vice-president of their company, after he had conferred by phone with a senior government official at NASA. When Jesse Moore was called back to the witness stand, having been exposed as either incompetent or untruthful, Chairman Rogers reassured him: despite having received a warning call " ... you decided that it was okay to go ahead – suppose that judgment was wrong. Nobody is going to blame anybody ... somebody has to make those decisions."

Unimpressed by that argument, Feynman asked more questions. Chairman Rogers tried to cut him off – twice – without success. Moore again put his foot in his mouth by explaining that, as there were two sets of "O-rings" reinforcing each other, it was safe to fly. In the past, he assured the commission, the secondary one held even when the first failed.

Astounded by the confession, Feynman protested: " ... you didn't expect ... (failure) on the first O-ring ... if the second O-ring gives just a little bit when the first one (burns through), that is a very much more serious circumstance, because now the flow has begun." At this point, another Commission member, Major General Donald J. Kutyna, who had directed space shuttle operations for the Department of Defense, interjected: "Once it got a path, then it burns like an acetylene torch."

It was Friday afternoon, Chairman Rogers hastily adjourned the session. On the weekend, Feynman contacted his former student, William R. Graham, the Acting Director of NASA, who remembered Feynman's lectures at Caltech 30 years before as "the best course I ever took." Graham arranged briefings with technical staff at NASA headquarters. Sitting among engineers, Feynman heard a litany of frustration with the bureaucracy; especially centering upon dissatisfaction with the O-rings. No tests had ever been run to learn how quickly the rubber bands should expand during blast - off heat. Nor was it known how quickly they should contract in sub-zero space chill, as the metal rocket-sections contracted. After previous flights, segments of the 37-foot O-ring had shown serious deterioration. When Feynman asked how this problem was then addressed, he was told that a computer was consulted on burn-resistance of the O-ring rubber. The A-bomb veteran thought this was ridiculous. When he asked if problems were reviewed between flights, the answer was that no review ever was held – until the next flight-readiness review. Feynman's bitter words: "If the seals leaked just a little and the flight was successful, it meant that the seal situation was not serious ... the seals could leak and it would be all right – it was no worse (a risk) than the time before."

So fear of failure was known – five months before the scheduled Challenger launch! Moreover, a written report, endorsed by NASA and Morton Thiokol engineers had flatly stated: "The lack of a good secondary seal in the field joint is most critical and ways to reduce joint rotation should be incorporated as soon as possible."

Nothing was done. Inertia: as damaging as purposeful negligence. After Feynman met with the NASA scientists and engineers, he learned that Rogers had tried to cancel the date with Graham. With commission colleague astronaut Sally Ride, Feynman requested transportation to the Kennedy Space Center to interview engineers. Graham approved. David Acheson at Kennedy approved. Rogers said no. A wall was being created.

Feynman next asked Graham for data related to performance of the rubber O-rings under flight conditions – resilience during vibration, heat, cold – in blast-off and in space. NASA produced data reporting on resilience over a period of hours in laboratory conditions, not fractions of a second during actual flight. Useless.

During the first days of the Inquiry, official reports to NASA bureaucracy by Richard Cook, the agency's budget analyst, were introduced. He had repeatedly – month after month – fingered the O-ring danger as a threat, not to human lives, but to budget!

Chairman Rogers, supposedly neutral, suddenly recalled his skills when he was chief prosecuting lawyer in the Department of Justice. He called NASA's Cook to the stand – and tore into him as a trouble-maker: "You had no reason to think that people who were weighing those considerations were not qualified ... You didn't feel that you were in a position (to) make decisions about what should be done with the space program? ... (you take) issue with the people who were highly qualified ... You didn't really mean to criticize your associates or people around you, did you?"

It was the one and only time Rogers interrogated a witness. It mattered not what was the substance of Cook's testimony. The charade was now crystal-clear:

whoever was negligent in the Challenger calamity, nobody would be held responsible. And possibly nothing would be done to prevent future ones. Whitewash, pure and simple.

Would Cook's testimony have any lasting effect? Defense Department and NASA contracts with Rockwell, Grumman, Martin-Marietta, Lockheed, and hundreds of their sub-contractors were – for political reasons – strategically spread over dozens of states and were vital to their economies – today, tomorrow and long into the future. For years, Congressmen from those districts had rubber-stamped every proposal for new NASA projects, however fanciful; then approved "supplementary budgets", for every cost over-run of every such project, however astronomic.

Chairman Rogers held a closed meeting of the Commission and deplored widespread reportage the day after the Cook revelation. Meddling by the press was, he said, "unpleasant, unfortunate ... there is no point in dwelling on the past." The public's notoriously short memory could be counted upon to fade, Feynman realized. The facts, the conclusions the Commission had reached would be buried in a file and produce no corrective action. What to do? He pondered this dilemma, dining alone at his hotel that evening. His eyes fell on a glass of ice water – and he knew what to do. He determined to jolt the nation into facing truth about the tragedy.

The next day was to be a public session with television cameras present. Early morning, Feynman drove in a taxi around wintry Washington in his light California clothing, hunting for a hardware store. When he found one, it wasn't yet open; he waited outside, shivering. When the store opened, he bought a screwdriver, a small C-clamp and a pair of pliers. Back at the hearing room, he unscrewed the "field joint" model that was used as an exhibit, and extracted a sample of the O-ring rubber.

The hearing opened with Lawrence Mulloy, NASA Solid-Rocket Project Manager, who went to great length, insisting that temperature had no relevance to the disaster. He showed charts, graphs, and used technical jargon.

Feynman had procured a pitcher of ice-water. Sitting next to him, General Kutyna pointed to the agenda printed on their "briefing book" and whispered "When he comes to this slide here, that's the right time ..." The moment arrived. Feynman tapped his microphone. Cameras swiveled and focused on him. He placed a section of O-ring rubber in the C-clamp, tightened the thumb-screw, and lowered it into the ice-water. Sixty seconds ticked by in silence. He reached into the pitcher, pulled out the C-clamp, unscrewed it, extracted the rubber from the inside of the clamp. The rubber retained the curled shape of the clamp. Holding it up for the cameras, Feynman said: "Mr. Mulloy, I took this stuff that I got out of your (O-ring) seal and I put it in ice water ... when you put some pressure on it ... and then undo it, it doesn't stretch back. It stays the same dimension (as in the clamp) ... there is no resilience in this particular material when it is at a temperature of 32 degrees." The temperature on the pad the morning of the launch, of course, was known to be much lower.

Chairman Rogers prevented Mulloy from responding. "That is a matter we will consider, of course, at length ... in the session that we will hold on the weather." He called the next witness, and hurriedly went forward with the agenda as though the bombshell had never dropped.

But it had. All America had seen on television that evening, the Feynman experiment in elementary physics – over and over again. All America now knew the cause of the tragedy – and the negligence that made it inevitable. No matter what happened thereafter, the truth was known. America's most influential newspaper, the New York Times concluded that the disaster had been inevitable: failure of even one O-ring would cause "loss of vehicle, mission and crew due to metal erosion, burn-through, and probably case bursting resulting in fire ... There is little question that flight safety has been and is being compromised by potential failure of the seals, and it is acknowledged that failure during launch would certainly be catastrophic." Editorial comment in other papers and television, nationwide, was blistering.

As his wife, Gweneth, had predicted, the "twelfth man," alone, continued his personal inquiry for months after the Commission's public hearings concluded. By interviewing engineers and technicians on the shop floors of NASA Space Centers in Houston, Florida, and Alabama, and the Morton Thiokol Company, Feynman learned that the rocket's engines had been operating at under 10% efficiency – 90% below their rated capacity. Because mechanical defects were found after each and every previous Challenger flight, but the mission wasn't destroyed – test standards for future flights were not tightened, but lowered! The National Research Council later analyzed Feynman's data, and estimated "probability of ... a catastrophe at 31 degrees Fahrenheit." The thermometer stood well below that on the morning of the fateful launch.

Feynman detailed these revelations – and more – in an engineer's analysis of NASA procedures, and asked that it be circulated among the other eleven members of the commission. He was told it had been done. But it would not be used because he had been out-voted. He phoned around, and learned that none of his colleagues had seen his summary. Instead, Rogers' staff writer drafted an "official" commission report that praised NASA as "a symbol of national pride ... NASA's spectacular achievements of the past (promise) impressive achievements to come."

Feynman wired Rogers that these fatuous words did not accord with his findings and he would publicize resignation from the Commission if his comments were not appended. Faced with this threat, the Establishment backed down and Feynman's personal warning was included:

"For purposes of self-preservation, NASA decided to exaggerate how economical the shuttle would be, to exaggerate how often it could fly, to exaggerate how safe it would be, to exaggerate the big scientific facts that would be discovered ..." He concluded ... "for a successful technology, reality must take precedence over public relations, for nature cannot be fooled."

Tended by his wife and sister, Feynman fought against the progress of his cancer until he succumbed late at night, February 14, 1988, two years after the Challenger fiasco.

The lesson he'd taught, however, lives on. His words: "Every time we talked to higher level managers, they kept saying they didn't know anything about the problems below them ... either they didn't know, in which case they should have known, or they did know, in which case they're lying to us." Feynman alerted scientists to their responsibility as citizens, not detached, passive onlookers. He alerted members of Congress and their constituents to the human fallibility of NASA people and the military-industrial complex that fed on NASA budgets.

When a heavy increase in funding for the multi-billion-dollar NASA Strategic Defense Initiative ("Star Wars") space project was proposed by the Administration, it was emasculated by an overwhelming chorus of disbelief at all levels: citizens, congresspersons, scientists, press.

A severely critical consensus was reached in a "Public Service Report" of the Federation of American Scientists, dated September, 1986:

"By far the most dramatic example of misplaced R & D resources in the Strategic Defense Initiative ... the chaotic and inefficient expansion of the SDI ... is draining large quantities of money and talent from military and civilian R & D efforts of far greater merit ... with adverse consequences for the deprived sectors and for the country as a whole.

"Research and development is essential to progress (in many peacetime industries) ... as well as in military preparedness. As a result, the vitality and productivity of our R & D enterprise affects ... our future international economic competitiveness, our prosperity as a nation, and our material and physical well-being as individuals. Our national security rests as much or more on these latter characteristics, as on our military capabilities."

Censure of slipshod NASA procedures appears in the body of the FAS Report lead editorial: "Oversophistication of ... design and undercompetence in manufacturing and testing have combined to help drive costs up and reliability down."

Jagdish Mehra's splendid biography of Feynman is entitled "the Beat of a Different Drummer." Would that more people – especially those in whom responsibility is reposed – have the courage to march to Feynman's drummer.

* *

This story is included in this compilation though –unlike all the others it is not centered on an individual's original work. It is all the more significant for that reason – demonstrating a sense of responsibility irrelevant to the mainstream of a supremely successful and highly honored career. "Beat of a Distant Drummer," the authorized biography of Feynman – and written with his help – devotes to the "Challenger" episode but six in 608 pages.

The structure of DNA
Was first unveiled by
ROSALIND FRANKLIN

"What every scientist knows ... is that the requirement for great success is great ambition ... for personal triumph over other men, not merely over nature. Science is a form of competitive and aggressive activity, a contest of man against man that provides knowledge as a side product ..."

These words appeared in the Chicago Sun-Times February 25, 1968, in a review of "The Double Helix." Author of the newly-published book was James Watson, Nobel Prize recipient for participating in the discovery of the helical structure of the deoxyribonucleic acid molecule (DNA). Review of the book was by Richard Lewontin, professor of biology at the University of Chicago Lewontin was upset by it. Others were, too, most especially Francis Crick, who, when he read Watson's manuscript before it was printed, Crick protested so strenuously against the "contemptible pack of damned nonsense" that Harvard University Press, which had previously committed to publication, withdrew from the contract. It was the first time in the Harvard editor's 21-year tenure that such a controversy resulted in the extraordinary rejection. The book was subsequently published by Atheneum Press, another prestigious house, to which Harvard editor Thomas J. Wilson transferred, manuscript in hand. It became an international best-seller not because of its account of a trailblazing achievement in science — though that in itself was exciting — but the scandal inadvertently exposed therein.

With the pace of a novel, step-by-step the book relates how the DNA mystery was deciphered. The narrative is peppered with gossip about real-life personages. And relates how, in constructing the double-helix theory on a foundation of work pioneered by another, the Nobel laureates won fame and fortune.

Their achievement was "the greatest discovery of the century" says Richard Dawkins of Cambridge University. It has been the basis for a worldwide concentration upon DNA medicine and biochemistry.

The work was pioneered by Rosalind Franklin, a researcher at London's King's College. It was her X-ray photography and the data developed as a corollary that revealed the unusual structure of the DNA molecule, and provided an answer to a great mystery – how genes replicate and convey information. From that point, the new science of molecular biology has emerged. She had assembled this information over two years of experimentation. Upon her findings, Watson and Francis Crick, his partner at Cambridge University, had erected their helical theory. Watson admits that

> "all my work has been getting other people to help me. If I have to use someone else to get the answer, I'll do it ... the most important thing in science is getting the answer, not showing that you've done it yourself ..."

That extreme pragmatism was spelled out 15 years after the double-helix "discovery," which had been announced in "Nature" magazine April, 1953. Their short essay failed to credit Franklin. Failed to credit, also, Linus Pauling, whose work on the structure of molecules and on hydrogen bonding led to DNA – or credit any of the earlier investigators in the field. The omissions were so offensive to the profession that every biologist who was approached by "Nature" to review and comment on the article, declined.

Watson and Crick's dubious ethics — co-opting somebody else's work without permission and without attribution, — are of course not unknown in science. (Antoine Lavoisier, discoverer of oxygen, failed to mention the assistance of Joseph Priestley, English chemist, who showed the Frenchman how to prepare the new gas 500% purer than ordinary air.) Ambition is an aggressive, and not always scrupulous, force. But it seems that more than ambition might have been involved.

Watson, in his book, reveals special interest in Franklin as a woman. He is displeased that "she did not emphasize her feminine qualities ... she ... might have been quite stunning had she taken even a mild interest in clothes ... there was never lipstick to contrast with her straight black hair ... her dresses showed all the imagination of English blue-stocking adolescents." He was further dismayed that she wore glasses. His assessment was very wide of the mark. Franklin had worked for four years in Paris. She was well-groomed, she designed and made her own clothing; she raised and lowered her hemlines as fashion dictated. She had an active social life in Paris, was popular with her French colleagues. She wore lipstick and didn't wear glasses. Jeremy Bernstein, author of "Experiencing Science," describes her as "a very attractive woman."

Watson's first view of Franklin was in November, 1951 when he attended a colloquium she headed at King's College at which she announced an important milestone in her DNA research. Test fibers were sensitive to humidity, she revealed; by controlling their moisture environment, she had developed X-ray photographs (identified as a "B" form) far more detailed than possible with ("A" form) dry fibers. The photographs revealed that phosphate sugars wrapped around the molecule's exterior, while sets of structural nucleic acid bases were buried inside, at right angles to the axis of the spiraled phosphate sugars.

So absorbed was Watson in his critique of Franklin's looks and dress that he and his partner continued with irrelevant experiments in their DNA research for 15 months as they failed to incorporate Franklin's discovery. "There were long intervals," Watson admits, "in which we were stuck ... just drinking coffee ... wondering why we could not think of the right answer." At one point, Crick and Watson showed Franklin a possible model of the molecule based on Watson's inaccurate remembrance of her lecture. Franklin – almost contemptuously – dismissed their formulation as containing only 10 percent of the water it should have.

In his book, Watson admits that he was a newcomer to DNA research whereas King's College scientists had been concentrating thereon for five years and were Britain's acknowledged leaders in the field. In fact, their chief, Sir Lawrence Bragg, aware that the duo was ill-equipped for crystallographic research, told them to avoid DNA work – "do not rock the boat," Watson recalled.

The Cambridge pair's failure to be aware of information that she had publicly disseminated left an unfortunate impression with Franklin. She saw model-building as a lazy approach to DNA research, avoiding the hard work required by accumulation of experimental crystallographic data — the procedure to which she was devoted. (Watson himself, in his book, opines that model-building, in her eyes, was an "easy resort of slackers who wanted to avoid the hard work necessitated by an honest scientific career.") Says Evelyn Fox Keller, of the Massachusetts Institute of Technology, "She really believed in doing the work for its own sake. She just loved science ... She was a woman of extraordinary integrity ..."

Disdain colored Franklin's attitude thereafter. In contrast, Crick confesses, "we always used to adopt ... a patronizing attitude toward her." An odd admission, considering that she had mastered an advanced research technique of which the Cambridge pair was almost completely innocent. Doubly odd in that Watson, particularly, was weak in chemistry, wherein her numerous publications had demonstrated superior strengths. As a student at the University of Illinois — he admits in his book — he tried to avoid taking physics and chemistry classes.

The partners' workplace in Cambridge was the Cavendish Laboratory, world-renowned as England's premier science center — except for DNA research. Its chief, Sir Lawrence Bragg, was a Nobel laureate who had pioneered crystallographic photography decades earlier. When he learned that Crick and Watson had strayed into misdirected DNA experiments, he termed it "the biggest fiasco" he had ever been associated with. He ordered the pair to quit work on a subject for which they manifestly were not equipped.

For five years, DNA research at King's College had been centered in the person of Maurice Wilkins. He was chief assistant to John Randall who headed the biophysics lab. Wilkins had previously worked for Randall as a graduate student before the war. During the war, he had worked on America's Manhattan Project. Since the war, he had concentrated on the DNA enigma at King's. Accordingly, he considered himself to be the senior person on Randall's staff, entitled to share whatever knowledge was being developed by others in the department.

Franklin didn't see it that way. Happy in Paris, she had been lured back to London by Randall, "to investigate the structure of certain biological fibres in which we are interested ... This means that as far as the experimental X-ray effort is concerned, there will be at the moment only yourself and (Raymond) Gosling," a graduate student. It would seem that Randall failed to inform Wilkins of Franklin's de-facto autonomy, with authorization to conduct, independently, DNA research. As Watson understands the relationship, "Sir John Randall – her boss – thought that really she was better trained than Wilkins to carry off the x-ray work on DNA ... although Wilkins had started it ... very soon they got on each other's nerves ..." The ground was prepared for a tragic clash of personalities.

Soon after Franklin's arrival in London, Wilkins had consulted her on his failure to dehydrate a DNA specimen. She provided the answer — a simple, almost elementary procedure. This experience may have affected her subsequent attitude toward the man. Clearly he was weak in basic principles of the discipline in which he worked. Whenever he requested information thereafter, she was abrupt, sometimes truculent.

She may also have been put off by his personality. He is described by colleagues as reticent, shy. He was ill-equipped to deal with Franklin who was outspoken, blunt, argumentative. This is normally a valuable trait in science. "It is one of the requirements," says Francis Crick. He goes on: "you must be perfectly candid, one might almost say rude ..." Raymond Gosling, Rosalind's assistant, says that she employed "a very sharp debating style of discussion ... you had to argue strongly ... Maurice would simply shut up," probably, as Watson surmises, "he was not a trained crystallographer ..." To colleague Aaron Klug, the hapless Wilkins confided "She scared the wits out of me." Watson: " ... it is hard to be successful in science unless you talk to your opponents. You have to get to know ... what their arguments ... are." Rosalind was excluded from the opportunity of dialogue.

Another major factor in their relations deserves even greater attention, one that she considered especially offensive. Wilkins didn't have a single woman in his research group. This anomaly was common at King's College, which had originally been a theological institution for males only. Women faculty members when Rosalind arrived were second-class citizens. They were barred from the faculty dining hall, relegated instead to the students' mess or forced to go out to lunch, whatever the weather. Rosalind believed she had been hired to create and run her own crystallographic operation, and with it, proceed independently with

DNA research. Notwithstanding the importance of the post, she found herself out of the loop — simply because she was a woman. This was unacceptable, particularly given her background.

From childhood, she had demanded of her three brothers equality in all matters. She shared their interest in mechanical toys and carpentry projects; she competed against them in athletic pursuits; she didn't play with dolls, shunned little-girl activities. Entering adolescence, she became resentful of the handicaps and constraints facing females in a male-dominated culture.

Her father made clear that a scientific career in Britain was almost impossible for women even to aspire toward. Very few females were employed professionally as scientists, engineers, or even technologists.

Rosalind had been born into a wealthy Jewish family. Upper-class Jewish females were supposed to earn their place in British society by immersing themselves in worthy civic projects. Her father proposed volunteer work — in which her mother and aunts were active. (He himself, though a busy lawyer, served as an unpaid teacher most of his adult life, at a Workman's College sponsored by a Christian Socialist organization.)

The family was not just wealthy, but in a certain sense, distinguished, descended from a famed rabbi. It had been resident in England since 1763. Mavericks, too. Though merchant-bankers, many members of the clan were strong supporters of the Labour Party. One was elected twice to Parliament but not seated, as he declined to take the oath of office on a New Testament bible, for him not a sacred text. (When the law was changed, his Christian constituents elected him a third time, and he was seated.) Another was an active proponent of women's rights, had gone to prison for physically attacking Winston Churchill, at that time an opponent of women's suffrage. An aunt was a trade union organizer and was elected to the London County Council, serving as chairperson at one time. Rosalind's great-grandfather, having matriculated at 13 on a mathematics scholarship, was graduated from University College in London with honors also in Classics. He became professor of political economy at the college — the first Jew in English history to hold such a position.

Rosalind inherited that same independence and drive. She rejected her father's advice to make a life in social service and decided, as early as her teens, on a career in science. She demonstrated an exceptional talent for it as a student at London's elite St. Paul's School for Girls. She applied to Cambridge to continue her education in science and was accepted. Her father, believing that path would lead to frustration and disappointment, objected strenuously, refusing to subsidize her tuition. His wife and his sister embarrassed him by pledging personal funds; outflanked, he retreated and thereafter paid tuition fees. (Rosalind confessed, later in life, that she had always nursed resentment for her father's resistance to her career ambitions.)

Virginia Woolf has testified that a gender-quota system was operative at Cambridge in those days; she estimated that "limited female admissions" were about one-tenth of male enrollments. What's more, female graduates were not awarded full degrees, as were men. Rosalind entered a women's college at Cambridge in 1938, at 17. A year later, when World War II erupted, her father, a World War I veteran, urged his daughter to quit Cambridge and "do "war work" as a volunteer. Fortunately, the government insisted that science students should remain in class.

She pursued an accelerated curriculum. During a skipped semester, she polished her French accent by living in Paris. This resulted from tutoring by a French woman whose Cambridge scholarship stipulated that she be available for intensive language study by students. Rosalind became fond of the woman; when she acquired a house to rent rooms to students, Rosalind moved in.

A year of post-grad study followed at Cambridge with a future Nobel laureate. A bumpy year. Her professor interpreted Rosalind's wilfulness as an aberration of the feminist movement; noting on the record that "she was not easy to collaborate with."

But Britain was at war and millions of men had departed the workplace. This created opportunities for females where they had never existed before. At 22, Franklin was employed as a chemist by the British Coal Utilization Research Association. Her industry-sponsored assignment was to devise methods whereby fossil fuels, the main source of energy in wartime Britain, could be utilized more efficiently. She studied — most of the time independently — the structural transformations of coal and charcoal when heated. She became an authority at creating high-strength carbon fibers. Hers was an extraordinary success: a young person, fresh out of school, accomplishing results of great value to the beleaguered nation. "She brought order into a field which had previously been in chaos," a colleague enthused.

From 1942 to 1945, she published five papers which are still referenced today. Her data were subsequently useful in the infant nuclear industry, too. She wrote her thesis during this time, and was awarded a Ph.D. in chemistry by Cambridge in 1945. It was now time for Rosalind to seek wider horizons in science. Her close friendship with the French woman in Cambridge, Adrienne Weill, resulted in offer of a job in Paris with The Laboratoire Central des Services Chimiques de l'Etat, a government institution. For nearly four years, she immersed herself in the refinements of crystallography — X-ray photography.

Crystallography early in 1946 was still a relatively new technique, employing an x-ray beam which passes through specimen matter to be scattered in a manner that projects a pattern on to a photographic plate. Diffraction of the X-rays delineates the position of the molecule's atoms; thereby capturing an image of its structure. Study of the intensity and angle of spots on an X-ray film thus helps to identify the distribution and placement of its atoms. It makes possible construction of a three-dimensional model that positions each atom in the molecule's composition. The results represent the molecule's anatomy in the round.

The crystallographic technique was becoming recognized as having direct application to studying the DNA composition of biological substances. Such evidence, it was hoped, would expose the DNA molecular inner structure. Toward that goal, a new generation of cameras and lenses was developed. Rosalind, accustomed to working with minute filaments in her previous carbon work — as tiny as a 200,000th of a millimeter — was superbly equipped to work with the new. highly sophisticated photographic equipment that had been developed.

This is what brought her to the attention of Sir John Randall; it was what led to the offer of a fellowship in his biophysics laboratory at King's College in London. This is the confirmation of the assignment she undertook there, excerpted from the first year's report of her work, submitted to Randall:

> "During 1950, M.H.F. Wilkins (had) succeeded in obtaining ... fine fibres from a specimen of nucleic acid ... and R.G. Gosling (produced) X-ray diagrams ... In January 1951, it was agreed that I should undertake, in collaboration with Gosling, a systematic X-ray investigation of these fibres."

The first eight months of that year had been invested in acquiring and assembling state-of-the-art equipment. During which time, she was exposed to an unpleasant discovery. She had not only worked during the war on a project of importance to the nation but also as an air-raid warden. Following which, she was accustomed to an amiable, gender-equal environment in Paris, where female scientists were honored, in the Marie and Eve Curie tradition. She had been surrounded by colleagues who liked her, included her in after-work picnics, parties, travels and entertainments. She made lifetime friends. At King's College, women were outsiders. They were excluded from after-work socializing at the nearby pub, where male colleagues exchanged information on their work of possible use to others. Only one other woman was employed in the entire biophysics department; no foreigners and no other Jews. Ignored as a scientist because she was a woman, she was denied opportunity to discuss work with peers. She was "out of the loop." So she kept to herself in the laboratory and organized her private life separate from her workplace. Resentful, she shunned sharing her findings with people who didn't respect her, and whom she didn't respect.

Most particularly this was true of Maurice Wilkins. He thought of himself as her senior. He believed that his seniority entitled him to get her help and to share in her findings. She thought otherwise. In comparison with the strong men in her family and the many good male friends she had known in wartime London and postwar Paris, Rosalind concluded that Wilkins was a cipher — and tact was not in her nature. Her assistant, Gosling, remembers: "She didn't suffer fools gladly ... You either had to be on the ball, or you were lost in any discussion about anything ..." Wilkins describes her, according to Watson, as "very fierce ... impossible as far as I was concerned to have a civil conversation." He complained of "superiority" in her attitude. Watson: "it was increasingly difficult to take Maurice's mind off his assistant, Rosalind Franklin, almost from the moment she arrived in Maurice's lab ... Maurice, a beginner in X-ray diffraction work, wanted some professional help ... (she) did not see the situation this way." Watson's use of the words "assistant" and "Maurice's lab" is significant. Franklin had been brought from Paris not to be anybody's assistant, but to be in sole charge of the laboratory's crystallographic work and – independently – to decipher the DNA mystery.

Rosalind's special skill in crystallography enabled her to isolate microscopically minute, gossamer filaments; and to hydrate them so as to photograph wet, as well as dry molecules. She would then apply a needle-sharp camera-beam focus upon them. The resulting pattern contained spots from diffraction at exceptionally small angles, which she was uniquely able to interpret.

Within 18 months after she started work at King's, she was able to focus a time-exposure X-ray beam on hydrated DNA fibers for 62 hours. The resultant picture clearly reflected its helical composition; the phosphate-sugars spiraled around the exterior of the molecule; the chemical bases were fixed inside. (Her photography was so precise that it is still reproduced in chemistry text-books a half-century later.) She made no effort to share this knowledge with Wilkins.

The hostility between him and Franklin provoked the man into strange behavior. He "borrowed" those photographs without her knowledge and duplicated them. What had taken her months of gradual refinement, he acquired in a few hours.

Soon after this episode, Watson visited King's. He barged into Franklin's laboratory without invitation. Watson: " ... as a American it just seemed natural to show up without appointments and letters, just to pop in. I was not constrained by good manners ... manners are just something that keep you from ... getting things done." He describes what happens next, in "The Double Helix":

> "Rosy was hardly able to control her temper, and her voice
> rose as she told me that the stupidity of my remarks would be
> obvious if I would ... look at her X-ray evidence ... I implied that

she was incompetent in interpreting X-ray pictures ... and hastily retreated ... Rosy ... firmly shutting the door."

Watson immediately described this encounter to Wilkins, probably volunteering an attitude that he later expressed in his book, to the effect that the proper place for a feminist scientist was in somebody else's laboratory.

"My encounter with Rosy opened up Maurice ...he could treat me almost as a fellow collaborator, Maurice went into the adjacent room to pick up a print of the new form ... the "B" structure ... mere inspection of ... (the) X-ray picture gave several of the vital helical parameters ... after only a few minutes' calculations, the number of chains in the molecule could be fixed ... Rosy had hit it right in wanting the bases in the center and the backbone outside ... Maurice told me he was now quite convinced she was correct ..."

Then followed the events that culminated in a Nobel Prize for Watson and the near-disappearance from science history of Franklin. Again, in his words from "The Double Helix":

"Afterward, in ... the train compartment, I sketched on the blank edge of my newspaper what I remembered of the B pattern. Bragg was in (the) office when I rushed in the next day to blurt out what I had learned .... Sir Lawrence ... urged me to get on with the job of building models."

Based entirely on what Watson had learned from Franklin's photographs, he and Crick skipped interim steps, especially the X-ray diffraction work for which they were not equipped. They proceeded directly to model-building ("Tinker-toy" play, as described by John Lear, the Science Editor of the "Saturday Review" in the March 16, 1968 issue). Watson summarizes: "roughly in six weeks the whole thing went through very fast." The tragedy then compounded.

Having recognized that "one cannot explain these clashes of personality," Sir John Randall decided to end them. He dictated that thereafter all DNA work at King's was to be under the control of Wilkins alone. He had Rosalind summarize in writing, all the data she had amassed in the preceding year, including the precise dimensions of the unit-cell skeleton inside the molecule. Randall next had Wilkins prepare an omnibus account of all work in the biophysics department for King's overseer body, the British Medical Research Council. It incorporated Franklin's statement. Wilkins drew attention that a "helical interpretation was very obvious in Franklin's fellowship report." (Reviewing the sequence of events, Robert L. Sinsheimer, chairman of biology at Caltech in America, opined in "Science and Engineering" (September 1968) that the double-helix "discovery would not have been long delayed" at King's.

Supervising body for all scientific research work in the country, the British Medical Research Council held its next meeting at King's. Randall gave each member of the Council a copy of Wilkins' summary report. Which included:

> "... when Rosalind Franklin began experimental work on DNA, she almost immediately obtained ... the first clear "B" patterns ... The helical interpretation was very obvious ... The best, and most helical-looking "B" pattern was obtained by Franklin in the first half of 1952 ..."

One of the members of the group was Max F. Perutz, who worked at the Cavendish Laboratory in Cambridge. He wrote in "Science" magazine (June 27, 1969):

"On 15 December 1952 we met in Randall's laboratory where he ... circulated the report ... Crick (later) heard about its existence ... and either he or Watson asked me if they could see it ... as a matter of courtesy, I should have asked Randall for permission to show it to Watson and Crick, but in 1953 I was inexperienced and casual in administrative matters ..."

Watson offers a slightly different account in his book: "As soon as Max saw the sections by Rosy and Maurice, he brought the report to Francis and me." He then calmly adds to the sequence of events an admission that the report contained "Rosy's precise measurements needed to check out" a helical construction. "Rosy did not give us her data ... and no one at King's realized that they were in our hands."

The chronology continues in Perutz' "Science" article:

> "a drawing of the ... patterns ...is also contained in a letter from Wilkins to Crick written before Christmas 1952. All this clearly shows that Wilkins disclosed many ... of the data obtained at King's ... "the (Franklin) report did bring the monoclinic symmetry of the unit cell home ... for the first time ... it suggested the existence of twofold symmetry axes running normal to the fiber axis, requiring the two chains of a double helical mode to run in opposite direction ..."

His last words in the "Science" article remind us, ironically, that Watson and Crick "could clearly have had this clue much earlier" if Watson had paid attention to Franklin's lecture in November 1951.

Within two months, the partners in Cambridge had constructed a model. All the molecule's atoms were arranged so that the diffraction pattern reflected the diameter of the helix along with other data procured from King's. In "The Double Helix" Watson crows: "Rosy had hit it right in wanting the bases in the center and the backbone outside."

Watson and Crick prepared their bombshell, the brief article for "Nature" magazine. The editor rushed it into print without prior peer review; it appeared before the end of April, 1953, and circulated internationally. The 1,000-word essay simply stated an hypothesis, offering no data, no proofs. The writers didn't provide explicit references to what had been accomplished previously as a way of clearly delineating what the authors were actually contributing. And certainly didn't mention that their double-helix theory was built upon others' work – apart from mentioning that they had "been stimulated by a knowledge of the general nature of the unpublished experimental results and ideas of M.H.F. Wilkins, Dr. R.E. Franklin and their co-workers" at King's. The essay gives greater prominence to a colleague at Cavendish, Dr. Jerry Donohue, for his "constant advice." Donohue, had explained the chemical structures, allowing Crick and Watson to see the pairing of complementary bases on the two DNA strands. This information – again furnished by another person, enabled them to deduce the structural basis of the genetic code. Critical information: it completed the double-helix theory, which was based, to begin, upon Franklin's results.

Also in the April 25 issue of "Nature," Franklin and Gosling supported the Cambridge theory, providing the photograph and further details. (Their paper had hastily been redrafted from an original version, which was to announce the helix construction themselves.) After which, Crick and Watson again appeared in the magazine, acknowledging the benefit of "X-ray evidence obtained by the workers at King's College, London, which gives qualitative support to our structure."

Why was Franklin so complaisant? Simply because she did not know at that time — nor ever in her short life — that her photograph and her data had been handed over to the pair at Cambridge. Years later, in his book, Watson quotes Wilkins as lamenting "If there had been anything like a normal situation here, I'd have asked her permission ... They could not have gone on to their model ... without the data developed here ... I blame myself."

With Franklin's consent, Randall transferred her fellowship stipend to Birkbeck College, there to explore the biochemistry of viruses.

The Birkbeck venue was primitive, lacking state-of-the-art facilities that she had created at King's. Her work spaces were separated by five flights of stairs, one being in the basement, the other in the attic. There were roof-leaks; Franklin worried daily about rain during the night. Buckets and basins cluttered the floor to catch drips. She affixed an open umbrella over her workbench when she left the lab at the end of each day. Nevertheless, her research proceeded swiftly and her reputation spread. Brilliant young scientists applied to work with her. Aaron Klug's work later led to reward of a Nobel Prize. (He had abandoned his own research project, on which he had worked for years, for the privilege of joining Franklin.)

The Birkbeck team set itself the mission of explaining the anomaly of virus particles, not themselves alive, being capable of growth and reproduction in other cells. Science historian Sharon Bertsch McGrayne ("Nobel Prize Women in Science") summarizes: "In Franklin's years at Birkbeck, her group outlined the general molecular structure of ... RNA-containing viruses and helped lay the foundation of structural virology ..."

It was learned that RNA plays key roles – as a messenger – in the transfer of genetic information from DNA to protein. It was learned that single-strand RNA can fold back on itself, forming double-strand regions.

During Franklin's tenure at Birkbeck, she published 17 papers on viruses and became recognized as one of the world's major figures in the emerging biomolecular science.

Captain of her own ship at Birkbeck, Rosalind was happy. A revealing clue to what may have been the personality problem at King's appears in Aaron Klug's comment that "in the lab she was actually quite tough ...(which) would have gone quite unremarked if she had been a man." Another colleague offers a similar clue: "She wouldn't put up with nonsense. She ... didn't indulge in speculation." The classic Greek proverb comes to mind: "Happiness is the exercise of vital powers along lines of excellence in a life affording them scope." The spirit was infectious; one assistant, later professor at a prestige American university, remembers "she affected all of us in a very deep way."

She had determined not to marry, not to dilute her total concentration upon science. She loved children, but could not face the likelihood of motherhood — too great a distraction from work. Unaware of this private resolution, the funding agency that had supported her fellowship, midway through her tenure at Birkbeck, expressed misgivings about their support of any project that has a woman directing it. Absurd, but her international reputation by that time impelled the U.S.Public Health Service to award her a three-year grant to continue her work, uninterrupted.

She didn't complete the term of the grant. Suffering from ovarian cancer for years, and in severe pain, she succumbed on April 16, 1958. She was 37, working till the end on poliomyelitis, a dangerous, infectious virus. Her boss at Birkbeck eulogized in "Nature:"

> "As a scientist Miss Franklin was distinguished by extreme clarity and perfection in everything she undertook. Her photographs are among the most beautiful X-ray photographs of any substance ever taken ... She did nearly all this work with her own hands. At the same time she proved to be an admirable director of a research team and inspired those who worked with her to reach the same high standard."

In 1962, Watson and Crick were awarded the Nobel Prize for Medicine or Physiology, in recognition that they had birthed a new science, molecular genetics, while Perutz and Wilkins received the Nobel Prize for Chemistry that same year, for their pioneering work in protein crystallography. It is believed that Randall had insisted on Wilkins being included in the group, so that work at King's — Franklin's work — would be recognized in proxy. (Nobel laureates must be alive when the prizes are awarded.) The acceptance speeches in Stockholm named nearly 100 scientists whose work was important to their breakthrough. Franklin was not mentioned once.

Six years later, "The Double Helix" was published, disarmingly cheerful, chatty, candid. And unsparing in its dislike and deprecation of Rosalind Franklin, described as a pathologically difficult, stubborn, belligerent personality. Reaction was so outraged in the world of science that Watson wrote, and published, an epilogue, admitting that he had been wrong about her. (As he didn't retract a single word of his original abuse, why the epilogue?) Interviewed by Anne Sayers, author of a splendid book, "Rosalind Franklin and DNA," Francis Crick voiced disgust: "Jim doesn't know what he is talking about ... He never understood what (Rosalind) was doing, he simply didn't know enough."

Watson's credo is worth repeating: "the most important thing ... is getting the answer." He frankly admitted the process that led to the double helix discovery, in a speech delivered at Harvard University in September 1999: "Francis and I basically stole the structure from the people at King's. I was shown Rosalind Franklin's X-ray photograph ... and a month later we had the structure; ... Wilkins should never have shown me the thing. I didn't go into the drawer and steal it. It was shown to me, and I was told the dimensions ... the Franklin photograph was the key event ... it mobilized us back into action ... we ... were able to say with finality that it was right because ... that came with Rosalind's X-ray work ... proof it was right ..."

Anthony Serafini, in his biography of Linus Pauling, twice a Nobel laureate and an also-ran in the DNA competition, comments: "There are so many actual and possible degrees of unethical behavior that it is difficult to draw the line. Sometimes, of course, the case is clear, as when James Watson made use of Rosalind Franklin's data without crediting her ... certainly Watson and Crick would not have gotten the Nobel Prize if they had not stolen her data."

Jeremy Bernstein in "Experiencing Science," heads his chapter devoted to Rosalind Franklin "A Sorrow and a Pity."

Naming of a major new campus, the "Franklin-Wilkins Building" – and placing Franklin's name ahead of the Nobel Laureate's – at long last is an acknowledgement by King's College of Rosalind's primary role in "the discovery of the structure of DNA."

GALILEO GALILEI
Hero or Craven?

Rare, very rare, is genius not manifested early in life. Galileo Galilei was 45 before he himself became aware that greatness awaited.

For 17 years, he had been a mathematics teacher at the University of Padua, comfortable and secure. At no time, had he expressed dissatisfaction. Had he not taken a ferry ride one day in 1609 across the vast lagoon that separated the mainland of Italy from the island of Venice, he might well have ended his days in Padua teaching math to young men. He was a good teacher, much respected by his pupils, some of whom kept in touch when their school careers ended.

Such a one was Jacques Badovere. He had been intrigued, on a business trip to Holland, with a curious novelty souvenir he'd bought at a carnival. It was a spyglass. Squint into one end, and the scene at the other appeared enlarged because of an ingenious set of lenses inside the tube.

On return to Italy, Jacques tried – and failed – to interest the Doge of Venice in buying his toy and developing it into an instrument useful to the redoubted Venetian Navy. It was too primitive. Maybe, if it were improved in looks and power?

He wrote to his old math professor, still in Padua. Not too far for a day trip. In Venice, Badovere showed his spyglass to Galileo — and history lurched out of its rut into a new direction. Though he had not previously shown an interest in optics, Galileo quickly understood that placement of a convex lens at one end of a spyglass tube, and a concave lens at the other, enlarged distant objects. He hurried back to Padua and started to tinker.

Six weeks later, Galileo Galilei demonstrated an eight-power telescope; 10 weeks after than, a 20-power model; in January, 1610, a 30-power instrument. It should be stressed that Galileo was not an astronomer – he was a mathematician, hitherto fully absorbed in mechanical physics.

Born eldest of seven siblings, he had enrolled at the monastery of Santa Maria, and briefly belonged to the Order. When he dropped out, studies continued at the University of Pisa. Medicine was his goal – but again he dropped out. From the beginning, he'd been studying mathematics privately, and soon began to teach it.

After four years, he applied for appointment to the chair in mathematics at the prestigious University of Bologna – and was passed over. He tried again at Pisa, and this time succeeded.

Like many independent thinkers during the Renaissance, he resented the knowledge strangle-hold of Aristotle, revered as the fount of all wisdom. During the nearly 2,000 years that had passed after his death, the Greek philosopher's axioms on all matters scientific were ironclad. For instance, Aristotle had ordained that bodies move with velocities proportionate to their weights. Galileo asked why, then, do chandeliers in the cathedral – of different sizes and weights, sway in identical arcs when set in motion by vibrations from the organ? He undertook the famous experiment by dropping both a large and a small stone from the top of the leaning tower at Pisa at the same time. They hit the ground simultaneously, of course. He exploited this small victory: derided colleagues on the university faculty who persisted in their fealty to Aristotle. He satirized their devotion to tradition – even the wearing of academic robes. Those familiar with the ways of academe will not be surprised that his contract at Pisa was not renewed at the end of its three-year term.

Unemployment then – as now – is a desperate condition. Job opportunities in his part of Italy were nil. Venice was then at the height of its prosperity. Venetian ships, navigators, and seamanship were without peer, funneling wealth from Aegean commerce into the lagoon. The chair in mathematics was vacant at the University of Padua, within the Venetian Republic. Galileo was hired. Busy years followed. He improved an existing instrument into a "geometrical military compass." It was capable of addition, subtraction, multiplication, division, square root, cube root, logarithms, measurement of densities and calculation of scales for the construction of fortifications. Valuable to Venice, valuable to every warlord and brigand in Italy, it became a best-seller. The teacher soon had an assembly-line to supervise.

Life continued placidly. Galileo was more or less satisfied with his lot. He was well-treated at Padua (isn't every professor who brings in a steady flow of outside revenue to his institution!).

But with telescope in hand, he aimed it at the heavens – and discovered what no man had ever seen before. The surface of the moon. It was amazing; it was intoxicating. It awakened in the mathematician an instinct toward poetry. "It is a very beautiful thing," he rhapsodized " ... rough and uneven ... just like the earth's surface with huge prominences, deep valleys and chasms ... lofty mountains ... (and) enormous peaks bathed in sunlight before the boundary of light and shadow reaches halfway across ..."

He expressed these enthusiasms in a small book that is still a pleasure to read. "The Starry Messenger," published March 12, 1610, promised a later, larger book, in which he would "by a multitude of arguments and experiences ... prove the earth to be a wandering body ... (with) an infinitude of arguments drawn from nature." The book caused a sensation all over Europe, and required multiple re-printings. It is still important.

Almost overnight, Galileo became famous. It boosted his own ego – which ever after led him into trouble.

His Padua appointment was extended "for life" and his salary doubled. But new horizons beckoned. Florence, his native city, was where the "Renaissance" had been born and was yeasty with creativity. An exuberant revolution against medievalism, it focused upon secular interests and scientific research, the flowering of painting, poetry, sculpture and literature, architecture, logic, and music.

During summer vacations, he returned to his family home there and tutored a lad in mathematics. That lad became Grand Duke of Tuscany, Cosimo de Medici. Galileo wrote into "Starry Messenger" a new term for astronomy: "Medicean Stars." He dedicated the volume to Cosimo. "Behold four stars reserved to bear your famous name ... moving about noble Jupiter ... with marvelous velocity ... executing mighty revolutions every dozen years about the center of the universe; ... I assign to these new planets your Highness' famous name ... Accept ... most clement Prince, this gentle glory reserved by the stars for you. May you long enjoy those blessings which are sent to you ... from God, their Maker and their Governor."

Grand Dukes are not immune to Flattery. Cosimo was, in addition heavily endowed with curiosity – about alchemy, magic, cabalism, all kinds of esoteric "sciences." Galileo was appointed as "First Mathematician" at Cosimo's University of Pisa; and "First Philosopher" to Cosimo himself. He had a large salary and few duties other than occasionally dazzling the duke's guests with lectures and demonstrations on astronomy.

Set for life? It wasn't to be. In his beautiful book "Saints and Sinners" (Yale University Press), Eamon Duffy explains: "The Republic of Venice was an Italian Catholic state which fiercely guarded (its) ... independence of the papacy ... It had Protestant mercantile communities within its territory ... Venice was devout and orthodox, but it policed its own orthodoxy ..." Pope Paul V actually excommunicated the entire Senate of the Republic, placed Venice under strict interdiction. No sacraments could be celebrated, Mass could not be sung in the churches, newborns could not be baptized, funeral rites could not include a priest.

So, though Venice remained Catholic - but was hostile to influence from Rome - the reverse applied in Tuscany. Clergy of Florence and Siena were keenly sensitive to Vatican politics, where an ultra-conservative faction of Jesuits was in the ascendance. The game of politics was an absorbing interest. Careers rose – or fell – as the game was played.

Like Leonardo da Vinci, the century before, who toadied to the Milanese dictator, Sforza, Galileo curried favor among the power elite in Rome. He entertained aristocratic personages and the princes of the church with his telescopes; many personally became interested in astronomy. (Who doesn't play with a new toy the morning after Christmas?)

He was riding high, with new friends at the power pinnacle of society – including Cardinal Maffeo Barberini, soon to become pope. In fact, Barberini wrote an ode in Latin, lauding Galileo's discovery of sun spots.

He curried favor not only for himself, but for the entire Florentine court. The Medici had vaulted him from obscurity to stardom; they provided transportation to and from Rome; paid a handsome stipend to cover his expenses there (where he lived in the Tuscan ambassador's palace) and guided him into the paths of power. Little wonder his growing intellectual arrogance.

On December 12, 1613, Cosimo Boscaglia – one of the men at the University of Pisa whom Galileo had ridiculed 20 years before – saw a chance for revenge. The grand duke had died. His son, 19, inherited as grand duke. The dowager Grand Duchess Christina was very pious. Boscaglia whispered to her at a state dinner that "Starry Messenger" supported the Copernican theory that earth rotated around the sun – which, of course, was repudiation of Holy Scripture. Rumors spread. Having been alerted to the possibility of trouble, Galileo circulated a response in a letter. He argued that the Bible and science were separate but equal sources of truth. Science sprang from observation, assembly of facts, and exercise of reason – which were human faculties endowed by God. As nature – all of it – was created by God, it was impossible for science and Christianity to be in conflict. Holy Scriptures had been written simplistically so as to be understood by the uneducated masses. He explained that Nature works by laws that care nothing for the ability of men to understand them; sacred literature should be used only to teach those things which could not be learned by man's own reasoning powers. He disposed of the Joshua stumbling-block, the stand-still sun. The letter was so detailed, so weighty, that it took months for Galileo's enemies to regroup.

Florence was the wrong battlefield; Galileo's prestige in his native city protected him. So the cabal sent a copy of Galileo's letter to Rome and denounced him to the Holy Office – the Inquisition.

Here enters one of the most respected, powerful, controversial figures in the history of the Roman Church, Jesuit Roberto Bellarmine. Still a towering figure in Catholic history – to this day – he was canonized as a saint three centuries after these events. He was then at the peak of his powers; the leading Catholic authority on doctrinal debates with Lutheranism. The pope said of him "he has not his equal for learning in the Church of God."

He crossed paths with Galileo at 73. He had been appointed, at 34, Professor of Controversial Theology at the Jesuit Collegio Romano, the most important educational institution in the Catholic world. Elevated to cardinal soon after, his influence grew. As students of power well know, every board of directors comprises many members, each of whom have one vote. But one member's vote carries more weight than all the others put together. Such a man was Bellarmine.

His position on doctrine had not changed since first enunciated: "the perversity of heretics is as much worse than all other afflictions as the dreadful and fearful plague is worse than the more common diseases." He saw dissidents as a "most crafty enemy;" it was his mission to lead "in the ongoing war against the foe of humanity itself."

He pointed out that study of the heavens by Copernican principles was one thing; but to conclude from there that the earth moves is a dangerous notion, injurious to holy faith. "The Church (cannot) tolerate that the Scriptures should be interpreted in a manner contrary to that of the Holy Fathers ..."

So he had laid down the law: "the Prophetic and Apostolic books ... are the true word of God and the certain and fixed rule of faith." There was no room for interpretation or debate, he made clear, because the final and sole "judge of the true sense of Scripture and of all controversies in the Church ... is the Pope ..." To foreclose any loophole-hunting, he wrote " ... the church absolutely cannot err ... (in matters) which it proposes us to believe or to do, whether they are set down in Scriptures or not."

This had been Bellarmine's conviction since early in his career: Authority must establish the essence of revelation, not human reasoning. As the historian of science, Richard Westfall, has written, "with all eternity stretching before him," the cardinal "could not understand why some men thought that the physical structure of the universe was so important."

> (Says Westfall, "nothing illustrates the reverence in which he was held by Roman Catholics better than the scenes, incredible to the 20th century, of cardinals and other prelates, touching and kissing his (dying) body, catching his blood in their handkerchiefs when leeches were applied to him, they refused to let the poor man die in peace. When death finally delivered him, they plundered ... relics, even ripping the clothes off his body ... As he lay in state (after death) the mob ... nearly tore the body apart ...")

Investigation of the charges against Galileo continued a full year. Galileo solicited support from church hierarchy whom he knew to be fond of astronomy and friendly to him. Most avoided "getting involved," but one added a strong hint. Galileo should write theoretically about mathematics, but "keep out of the sacristy!"

Though there is a general agreement among scholars that Bellarmine was immovable on the principles he stated so firmly, there is equal agreement that he held no personal animus toward Galileo. Their meeting must have been amiable, not threatening, because the accused penitent shrugged off the well-meant warning to be prudent. He was annoyed that the Church disdained even to answer the arguments of science.

Bellarmine's investigation rendered judgment February 26, 1616: Galileo's position was "as much a heresy (as) to say that Abraham had not two sons and

Jacob twelve and that Christ was not born of a virgin." The cardinal banned Galileo from further discussion of the sun-centered theory of the solar system. He prohibited public dissent from orthodox theology. He privately urged the scientist thereafter to maintain a low profile.

Galileo chose to interpret the gag-rule as applying only to printed heresies. He published nothing for seven years. But privately, he continued to teach and discuss radical and provocative ideas with friends. Fingers crossed, he explained heliocentricism as a shortcut to making calculations, not denial of doctrine. But his pussy-footing prudence at last backfired. A well-respected Jesuit professor of mathematics at the Collegio Romano, Father Orazio Grassi, delivered a lecture on the phenomenon of comets. Grassi dragged in Galileo's name and charged that, though he no longer put his heliocentric convictions into print, he continued to be a closet Copernican.

At the University of Pisa, another opponent, Ludovico delle Colombe, continued the renewed campaign against Galileo. Galileo rebutted with a 350-page essay, "The Assayer." In it, he abused his rival: "There is no point in undertaking to refute someone who is so ignorant it would require a huge volume to refute his stupidities." He went on, and refuted them one by one. From that point, he derided his enemies – among them the Jesuits – as members of the "League of Pigeons." (Colombe in Italian vernacular means 'dove' or pigeon.") "The Assayer" text was literary, eloquent, acerbic. The brash young priest was likened to a duck, "incapable of following the flight of eagles."

Galileo showed the essay to an old friend, Pope Urban VII, the former Cardinal Maffeo Barberini, who chuckled heartily at the wit. Word of the pope's pleasure got around the Vatican. The chief Censor, Father Niccolo Riccardi, who had been introduced to astronomy by Galileo and still was a fervent admirer, had no hesitation in authorizing publication of "The Assayer."

"The Assayer" was a repeat of the Pisa blunder – a great success and a gross mistake. It provoked Father Grassi's weightier colleagues at the college into a vendetta. Father Christopher Grienberger, Grassi's senior, and himself a respected scientist, lamented "If Galileo had known how to keep the affection of the fathers of this college, he would live in glory before the world … (to) write at his pleasure about any subject, even … the movements of the earth."

But it was too late; the scientist had gone too far. The Inquisitors were pressed to take action. Galileo's "heresies" were now so overwhelmingly evident and explicit that the "trial" would have gone badly for him. Pope Urban VII swept aside the charges on condition that Galileo thereafter would observe the "gag rule" of 1616 – no exceptions.

The second interregnum in the scientist's public life lasted fourteen years. He quietly continued research into gravity, inertia, and momentum – but didn't draw public attention. He was working on a new book. For two years, he wandered the corridors of the Vatican, seeking permission to have it printed. To be safe, he phrased its beginning and ending as conjecture, not scientific fact. This strategy was seemingly successful: the censorship office approved. In February, 1632, Galileo's "Dialogue Concerning the Two Principles of the World" was printed. It was a colossal triumph throughout Europe almost instantly. (The young Isaac Newton owned a copy.) Though "masked," it was a powerfully argued justification of heliocentrism. In the form of a Socratic discussion, the Copernican description of the Earth in movement was demonstrated as consistent with the general laws of motion. Every principle of physics articulated by Galileo's dialectic proxy was supported with ironclad proofs. Galileo displaced the Earth from the center of the universe; there "we shall rather find the Sun." Using the principles of inertia and the relativity of motion, he established the canon of centrifugal force. These mechanical truths, provable by mathematics, governed all terrestrial movements; they were the same for celestial movements. Thus Galileo strengthened beyond attack Copernican theories of the planets in a system of heliocentrism.

Exposition in "The Assayer" of the traditional, Ptolemaic position was, not surprisingly, weak. But, however camouflaged by the pretext of a conjectural debate, fundamental church interpretations of Scripture were exposed to question. Galileo all but prohibited enemies from citing Scripture in countering his arguments. Authors of antiquity — upon whom, till then, all "authority" was based — were proved fallible. Particularly, he scorned the outmoded axioms of Aristotle, whose thoughts were "badly" and "erroneously" conceived. Worse, he quoted the pope's own words — written years before — in a fatuous argument advanced in the dialogue by an obvious simpleton. It was, said the pope, a one-time friend and patron, "an injury to religion ... and of a perverseness as bad as (he had ever) encountered."

Despite all the favor shown to Galileo in the past, and the promises made by him, he was once again the epicenter of an historic struggle. Urban and Bellarmine let the hapless scientist suffer for seven months. Galileo was now 70, in poor health, and in severe pain from inflammation of his overworked eyes.

At Inquisition trials, the accused was not permitted to know what charges he must defend against or be represented by a legal advisor. He was not permitted to cross-examine witnesses. Galileo was found guilty of echoing Copernican theories about the solar system. On April 27, 1633, Cardinal Bellarmine, now the Inquisitor General, came to the prison and demanded that the prisoner "confess his error" — or face torture. He confessed his error. Three days later, he was further forced to disavow everything he had ever written. He did so. On May 10, hauled before the tribunal yet again, he pleaded for clemency: "At the age of 70, I have been reduced by ten months of constant mental anxiety ... (let my) sufferings ... be considered as adequate punishment." The Inquisitors turned a deaf ear.

On June 21, they forced him to his knees, literally to his knees. Again and again, the terrified old man was made to recant. The panel of cardinals then pronounced Galileo guilty and prescribed punishment. All his books were banned and he was condemned to lifelong imprisonment.

Having humbled the feisty maverick — taught a lesson still remembered today — Urban VII finally commuted Galileo's imprisonment to exile under house arrest, in a small Tuscan village.

Lucky to escape with his life, the parolee should have been content to live out his remaining time quietly. He was skilled in art; played several musical instruments; loved gardening. Instead, he continued working. In 1633, yet another book — his last and most important — was ready. Again it was in the classic form of Greek debate: "Discourses and Demonstrations Concerning Two New Sciences." (The subject he'd promised in "Starry Messenger.")

With a wary eye on Rome, he sent the manuscript to Protestant Holland, where the papal Inquisition had no influence. To attract the broadest possible readership, he had written not in Latin, the language of scholars and priests, but in vernacular Italian.

The book is a summing up: truth can only be recognized when proof of ideas can be tested and thus established beyond any doubts. Not faith or the invocation of authority, but facts, only reproducible facts, supported by evidence, must be established.

Again, this valedictory opus was an instant triumph. Distinguished admirers from all over Europe called upon the old man, by now going blind. In July, 1637, he lost the sight of his right eye. Six months later, all vision was gone. He wailed "… this heaven, this earth, this universe, which I … had enlarged a hundred thousand times beyond the belief of the wise men of bygone ages, henceforward for me is shrunk into such small space as is filled by my own bodily sensations." He lingered five years in the dark and died January 8, 1642.

Courage has more than a single face. Only decades before, Giordano Bruno had shown one kind of courage, defending Copernican theory unto pain and death as the price of integrity. Was Galileo less courageous? To avoid being silenced — like Bruno — he jousted consistently against immensely powerful opponents, exhausting his strength.

Centuries later, the Vatican issued a sincere, graceful apology — and sponsored impressive scholarship to illuminate the crisis that has enmeshed science versus faith, ever since the time of Bruno, Bellarmine, and Galileo.

Galileo's stubborn swimming against a strong tide has inspired independent minds ever since. Descartes was in Rome at the time of Galileo's last humiliation. He thereafter trimmed his sail to avoid church disapproval of what he knew was lethal — his four principles of rational philosophy. Leibniz and Newton contributed calculus less than 50 years later.

Einstein summed up the practical results of staying alive — on thin ice: Galileo's methods represent "the dawn of modern physical thought."

Frederick Goldman

JOHN HARRISON
Unschooled, he succeeded
Where Galileo and Newton failed.

The King of England personally nudged the scales of justice to pay aged and ailing John Harrison money he had long been owed by His Majesty's Government for the invention that helped Britain's shipping to become the safest in the world. It was a calculating-clock, capable of organizing a mass of data to reckon a ship's longitude – hence its position – at every moment. Lacking that knowledge, even skilled navigators sometimes steered into shoals and rocky shores.

It was just such a mishap that inspired Parliament to offer an enormous prize to whomever could solve the problem that had doomed mariners since time immemorial. On October 21, 1707, Admiral Sir Clowdisley Shovel, aboard his flagship, was approached by a deckhand. He saluted and reported that his personal calculations positioned the fleet widely off from those of the navigation officer. They'd been sailing in rough weather for days. Absolute precision could not be expected, true. But that a lowly swabbie should contradict officers' judgment? Impossible! The man was summarily executed for mutiny and dumped overboard. The very next day, the fleet ran aground – miles off its course – on rocks rimming the Scilly Islands near Cornwall, western tip of England. Four of the five vessels sank; over 2,000 men died, including the admiral.

The catastrophe galvanized Parliament. A huge reward was offered – millions of dollars in today's value, enough to provide a family fortune for generations to come. The challenge: devise a navigational instrument that would provide (a) instant data about a ship's position by relating the exact time aboard, with the home-port time, at exactly the same moment – even if "home" were on the other side of the world; (b) the instrument must be impervious to heat, humidity, freezing cold, abrupt changes in temperature, water, wind, and corrosion; (c) it must be delicately balanced yet rock-solid when the vessel tumbled, rolled and pitched in mountainous seas.

Harrison designed and built, over many years, a sea-clock to solve that age-old problem; it enabled the Royal Navy to become dominant on the high seas. But the Royal Observatory withheld payment year after year, with pretexts and lies. John's son, William, finally wrote to the monarch. His father was in pitiable state; likely he would die soon. The king, himself a naval man, granted an audience to which William brought with him his own young son. Sixty years later, John Jr., in his memoirs, remembered the king's outrage: "These people have been cruelly treated," he sputtered, and pledged, "By God, Harrison, I will see you righted!"

Long before Harrison, geniuses such as Galileo and Sir Isaac Newton had tried to devise such an instrument – and they had failed. John Harrison, firstborn son of a village carpenter in Yorkshire, England, was not school - educated at all. So how come? He learned everything by doing. He learned his father's trade; learned to play music at church; taught himself to read and write; learned physics from a borrowed book; absorbed the laws of motion by writing every word in that book over and over, copying and labeling every diagram. Grammar, however, spelling, and punctuation – they remained mysteries. When he wrote memoirs toward the end of his life, the first sentence ran over two dozen pages.

From age 19, Harrison built pendulum clocks entirely of oak and boxwood – not to order, but to test his skills. Report of these remarkable timepieces reached the ears of the local squire, Sir Charles Pelham. In 1720, the magnate hired him to build a clock tower over stables at his estate, Brocklesby Park. It took two years; the clock has since run for 275 years, correctly. Harrison used for all the moving parts of the mechanism lignum vitae wood, which oozes grease. The works have never needed lubrication.

No lubrication! Before Harrison, the very idea was unimaginable. It was axiomatic that oil is essential to counter friction in moving parts. The unstable acidity and viscosity of oil is one of the factors that prevented success in all previous attempts to devise a precision maritime instrument. Petroleum thins and thickens with fluctuations in temperature. Sea voyages involve transitions in climate. The moving parts of a lubricated clock will run faster or slower, as the thermometer rises and falls.

The metals in pendulums stretch and shrink with temperature changes. Expansion and contraction vary with each different metal and with its length and thickness. So he devised a "gridiron pendulum," with the bob suspended by a series of parallel brass and steel rods, alternatively arranged. The downward expansion of steel, by endless experimentation, was compensated into absolute equilibrium by the upward expansion of brass! He then invented a "secondary spring" to keep the timepiece ticking correctly, even when briefly it was being wound.

He then taught himself astronomy. Rotation of the earth in relation to stars' fixed positions creates the illusion of a "transit" nightly, precisely 3 minutes and 56 seconds earlier than the preceding night. Harrison tested his clocks against solar time each night, and continued to make adjustments in the "works" until they never erred as little as one second in an entire month!

With each innovative accomplishment at this workbench, Harrison's confidence grew. By the time he reached 40, he knew he could fulfill the prize-offer requirement for a maritime computer.

It was 1730, and he had never before been to London. The Board of Longitude had no office address. (Proposals from competitors for the prize had, till then, been so absurd as not to warrant summoning members to a meeting.) However, Harrison knew the name of one member of the Board, and to him he hastened.

The aged Sir Edmond Halley – yes, the man after whom the comet was named – headquartered at the Royal Observatory in Greenwich. He understood perfectly Harrison's theory and was unperturbed by the clockmaker's rustic vocabulary. But he knew that the other commissioners – astronomers, mathematicians, naval officers – would not. So he sent Harrison to see the most respected horologist in England, George Graham, in hopes that the clockmakers' ideas could somehow be made clear to conventional minds.

It doesn't happen often, but when it does, a first-time meeting can be so felicitous as to make history. After 10 hours of conversation, the two men became friends for life. Harrison later recalled: "Mr. Graham began as I thought very roughly with me, and the which had like to have occasioned me to become rough too; but however we got the ice broke ... and indeed he became at the last vastly surprised at the thoughts or methods I had taken." Graham went further, he lent the village carpenter enough money to get started on building a model.

That took five years. When finished, the clock weighed 75 pounds, and was housed in a case four feet square. George Graham arranged for a demonstration before the Royal Society. The distinguished members were delighted with its performance, attesting

> John Harrison, having with great labour and expense, contrived and executed a machine for measuring time at sea upon such Principle, as seems to us to Promise a great and sufficient degree of Exactness. We are of the Opinion, it highly deserves Public encouragement, in order to be a thorough Tryal and Improvement of the several Contrivances, for preventing those Irregularityes in time, that naturally arise from the different degrees of Heat and Cold, a moyst and dry Temperature of the Air, and the Various Agitations of the Ship.

Harrison was ordered to transport the wondrous device to the H.M. Centurion, which embarked on a sea-test voyage to Lisbon, May 4, 1736. He was provided with a letter hand-signed by Sir Charles Wager, First Lord of the Admiralty, which explained that the voyage was to determine whether

> The Instrument ... will succeed at sea ... (Harrison) is said by those who know him best, to be a very ingenious and sober

Man, and capable of finding out something more than he has already ... be as kind at him as you can.

In these gracious words is encapsulated the essence of the need for life-long education – "capable of finding out something more than (one) has already." In similar vein, the captain , wrote back:

> " ... I find (Harrison) to be a very sober, and very industrious, and withal a very modest Man, so that my good wishes can't but attend him; but the Difficulty of Measuring time truly, where so many unequal Shocks and Motions, stand in Opposition to it, gives me concern ... but Sir, I will do him all the Good, and give him all the Help, that is in my power ... he shall be well-treated ..."

In Lisbon, Harrison and his timepiece were put on H.M.S. "Orford" for return to England. Reminder: the vessels were tiny, and depended entirely upon the vagaries of wind and weather for forward progress. The outbound sailing took one week; the return trip took a month – in the teeth of recurrent storms. When land was sighted finally, the ship navigator's calculations reckoned the ship to be near Dartmouth in England, on the south coast of Devon. Harrison demurred, insisted the ship was approaching the Penzance peninsula, 60 miles west. Harrison was right. Captain Wills filed an affidavit June 24, 1737, admitting that his computations had been in error, Harrison's accurate. The story should have had a happy end when, one week later, the Board of Longitude met for the first time in its 23 years of existence. It had the power to grant the 20,000 pounds prize pledged by Parliament. Six of eight Commissioners were on record as favoring Harrison: Dr. Halley; Sir Charles Wager; Admiral Norris, who had examined the "Centurion's" log upon the ship's arrival in Lisbon; Dr. Robert Smith, Professor of Astronomy at Cambridge; and Dr. James Bradley, Professor of Astronomy at Oxford – both were members of the Royal Society and had signed the letter of endorsement; also Sir Hans Sloane, president of the Royal Society.

There was one other man at the meeting, who knew the instrument intimately – and knew there were defects: John Harrison.

During the voyages to and from Lisbon, errors of a few seconds in 24 hours had troubled him. He did not wish to deliver a "good enough" piece of work. So he promised he would return with a better clock – within two years. Even in an era when pride of craftsmanship was far more appreciated than it is today, Harrison's fanaticism was amazing. William Hogarth, a watch-face engraver early in his career, depicted him as the "longitude lunatic" in "Rake's Progress" (but privately admitted that Harrison's clock was "one of the most exquisite movements ever made.")

The Board had nothing to lose by granting the request. They had something to gain, though – and spelled it out in blunt detail: Harrison had to relinquish ownership rights to the instrument, though it had been created during five years using only his own and borrowed money! The inventor was not miffed, he had finally been recognized and, in a sense, encouraged to continue with his mission.

In January 1737, he appeared before the Board again. His second clock had been tested by the Royal Society, and endorsed as "sufficiently regular and exact, for finding the Longitude of a Ship within the nearest Limits proposed by the Parliament ..." Notwithstanding the Royal Society's accolade, Harrison once again expressed disappointment with his creation, once again asked for time to continue work. He was by then 48.

The race to invent a longitude instrument meanwhile was becoming crowded with entrants. An American, Thomas Godfrey, hit on the idea of using the moon as keystone to his longitude calculations. The same idea, almost at the same moment, was presented to the Royal Society by John Hadley, an Englishman. His device used paired mirrors, which reflected elevations of two heavenly bodies, and made possible measurement of space between them. Seamen called this instrument a quadrant. (Eventually, by extending the measurement arch and adding a telescope to the device, the name sextant was adopted.)

The moon-method had one major asset going for it. Mariners understood – had always understood – looking to the sky for guidance. They were comfortable with the complexities and intricate adjustments that had to be factored into measurements. They were flushed with new knowledge acquired at the Royal Observatory, which had but recently charted triple the number of stars than were known before. All they needed to complete lunar measurement of longitude were precise tables of distances between the moon and other heavenly bodies.

Necessity, it has been said, is the mother of invention. Tobias Mayer, a mapmaker in Nurenberg, Germany, delivered as his entry in the Longitude Sweepstakes, exactly what was needed in 1757. The Board, packed with sky gazers, was keenly interested in the potential this information offered. An interim award of 3,000 pounds was authorized.

Whereas, during the long trial-and-error process that is essential to original ideas, John Harrison received but 2,500 pounds interim payments. In 1759, he submitted his fourth clock, fruits of 19 years' work at his bench. (The previous model, H-3, had worked well, but true to form, the inventor had decided – just before its sea-test – that it was still too large.) In H-4, Harrison had miniaturized components in the works to an incredible degree. The entire casing now measured only five inches in diameter. H-4 did everything Parliament had demanded in the Act of 1714 – and more. Harrison was ecstatic: "... there is neither any other Mechanical or Mathematical thing in the World that is more beautiful ... and I heartily thank Almighty God that I have lived so long ... to complete it."

Less enthusiastic were the men who had to decide between it and the lunar method. They had the same reaction to technology as did aborigines upon seeing the first airplanes. Magical and mysterious! The jewel movement, anti-friction wheels, and hundreds of other parts were now invisible. The Board could not be expected to understand how the occult device worked; how it could be repaired when necessary. They were neither ignorant nor obtuse men. So, before we rush to judgment, we should appreciate what all historians preach: put events into the context of time and place. Precisely at a time that H-4 was submitted, astronomy was exploding with new knowledge.

The first-ever Astronomer Royal was famed John Flamsteed. His widow, Margaret, saw to the publication posthumously of her late husband's "Great Catalogue," his "Coelestic Brittanica" (1725) and finally, his magnificent "Atlas Coelestis" (1729). Another formidable woman was Caroline Herschel, who came to England from Hanover, Germany, in 1772, to be her brother's assistant, using telescopes up to 40 feet long. Their "sweeps" of the northern heavens vastly enlarged Charles Messier's list of 103 nebulae and clusters to over 4,000.

This was the same era when Tobias Mayer, in Nurenberg, developed the amazing lunar tables. It seemed to all that looked skyward, that lunar knowledge was on the verge of becoming as scientific as solar. Knowledge understandable not just to one or two unlettered men with a secret wrapped in a magic miniature package – like alchemists in the recent past – but to all educated men thereafter. And lunar charts cost a tiny fraction of a Harrison clock!

But Harrison father and son did not stand alone. The weight of the Royal Society required the Longitude Board to proceed with sea testing. In November 1761, son William sailed aboard the H.M.S. "Deptford" bound for Jamaica. H-4 sat snugly inside a box with four locks. Four locks! William had only one of the keys. When the case was opened daily to wind the watch, the process was observed by Dudley Digges and J. Seward, skipper and first lieutenant, respectively, of the vessel; and William Lyttleton, enroute to Jamaica to be Crown Governor.

Three months later, January 19, 1762, "Deptford" coasted to the dock at Port Royal, Jamaica. After 81 days at sea, H-4 was slow by five seconds.

The return voyage to England was a nightmare. It stormed much of the way. Seas were so heavy that the deck was often awash, knee-deep. Blankets were wrapped around H-4 to keep it dry, in the captain's cabin. When they were soaked, William wore them next to his skin, so body-heat would speed drying. Through it all, H-4 ticked cheerfully, and was slow less than two minutes when the ship reached home – well under the limits set by the Longitude Act.

But the rules of the game were then abruptly changed. When he was astronomer on staff at Oxford, James Bradley had been an enthusiastic supporter of Harrison. Now he was Astronomer Royal. Bradley had become facile in use of the lunar tables developed by the German, Tobias Mayer. William Harrison describes an encounter with him. "The Doctor seemed very much out of temper and in the greatest passion told (us) that if it had not been for ... (our) plaguey watch, Mr. Mayer and he should have shared Ten Thousand Pounds before now!" If William's words are taken at face value, Bradley had a conflict of interest.

Circumstantial evidence reaches a similar conclusion. The Longitude Board called for mathematicians to check the time-readings at the port of embarkation, Portsmouth, and the destination, Port Royal – the affidavits of the Board's own observers, on the spot in both places, was deemed inadequate! Bradley died suddenly, but his replacement, Nathaniel Bliss, went even farther afield. He threw out all proofs of the accuracy of Harrison timepieces, terming them sheer coincidence. He demanded another test voyage to the West Indies. Plus full disclosure of how the mechanisms worked, never before a requirement of the contract.

Huffily, the Harrisons refused, but were worn down after 18 months. William and H-4 boarded the H.M.S. "Tartar;" Captain Sir John Lindsay supervised the daily winding and monitored the review-checking upon arrival at the Barbadoes. Among the "monitors" was a face that William knew well, Bradley's protégé, Reverend Nevil Maskelyne – who had loudly and publicly already declaimed that the Lunar Method was the better way to go. Word of his boast was brought to Captain Lindsay and William Harrison. They disputed the right of Maskelyne to participate in judging - which resulted in a long delay and enmity between the men.

Months went by, but eventually the Board's mathematicians, who had checked astronomers' certifications of both the longitude at Portsmouth and longitude at the Barbadoes, reported "unanimously (we are) of the opinion that the said timekeeper has kept its time with sufficient correctness." "Sufficient correctness" was the reticent confirmation of a performance three times more accurate than called for.

All's well that ends well – right? Wrong. Harrison's opponents on the Board still had a rearguard action to fight. They proposed paying half the reward money upon delivery to them of all four Harrison timepieces. Plus all drawings, diagrams, notes and specifications for H-4. Plus two duplicates of the miracle watch – to prove that it was not a one-off fluke.

Nathaniel Bliss died, only to be replaced by 34-year-old Nevil Maskelyne. At the very first meeting of the Board he attended in his official capacity, he delivered a further setback to progress, demanding still another comprehensive review of the lunar versus the chronological measurement of longitude.

A new Act was voted in Parliament – 50 years after the original specifications for which the prize was offered. Aimed squarely at Harrison, it required him to undergo grilling by a delegation of six men, including Nevil Maskelyne. Harrison was forced to take apart H-4 – in their view – and explain the function of each of its hundreds of parts, including how each of his innovations interacted in a self-compensating scheme.

At the end, Harrison was told to put the watch back together and turn it over with all his handwritten drawings, diagrams, and descriptions – which Maskelyne appropriated. But that was not the end; the Astronomer Royal appeared at Harrison's door a few days later with a court summons and confiscated all three sea-clocks made before H-4. (One was dropped by Maskelyne's men exiting the house.) They were then placed on a flatbed, unsprung cart and bounced across the cobblestoned streets of London. The maker was never again to have them in his possession.

Transported to its new home in Greenwich, Harrison's H-4 reacted poorly. Was it because it was damaged during transport? Was it because it was fastened in direct sunlight next to a window as in a greenhouse? Was it the unfriendly hand that wound it daily – and noted in the log, "Mr. Harrison's watch cannot be depended upon to keep the Longitude within a degree ... in a (long) voyage ... nor to keep the longitude within half a degree for more than a few days ..." "Maskelyne admitted that a timekeeper such as Harrison's might be of use as a backup to lunar calculations – on the six days every month when the moon is too close to the sun; and on the 13 days per month when it shines beyond the horizon and is invisible to a navigator.

Unfazed, John Harrison grimly proceeded with the dictate that he build two replicas of H-4. He had no access to the original, or the notes and diagrams he had recorded over the years, confiscated along with the watch itself. Despite Maskelyne's poor report, the Board deemed H-4 too precious to leave England, so a twin was ordered from Larcum Kendall, a watchmaker who had participated in the six-day interrogation of Harrison. He had access to H-4 for guidance, so finished his work in 1770. Harrison took longer to complete the first of the two watches he'd been ordered to make – then took another two years to test it.

With eyesight failing, at 79, he felt unable to start work on the second of the two. Even if he had years more to live, it was obvious that further delays and digressions would be ordered. In despair, son William wrote to King George III , which is where this story began. The king, his science advisor and William Harrison tested H-5 at Windsor Castle, in the king's private observatory, to duplicate Maskelyne's claimed-unsatisfactory procedure with H-4. Every day at noon the men met. Three separate keys were used by them to unlock the clock-case for the winding ceremony. At the end of 10 weeks, in July, 1772, the clock was certified as accurate within one-third of one second per day.

Only then, payment was made, 10,000 pounds, not full amount owed under the original act – which Harrison never did get. Kendall's copy accompanied Captain James Cook on his three-year circumnavigation of the globe. The navigator lauded the Harrison invention as having "exceeded the expectations of its most zealous advocate ... (it) has been our faithful guide through all vicissitudes of climates ... (I) would not be doing justice to Mr. Harrison ... if I did not own that we have received very great assistance from this useful and valuable timepiece."

Kendall made another clock, which Cook took along with K-2 on his ill-fated 1776 repeat voyage. The instrument performed well and returned from that trip, but not Cook, who was murdered by Hawaiian natives in 1779.

Thomas Mudge, a watchmaker, had also been an "observer" at the star-chamber session in Harrison's house. Not commissioned by the Board to do so, he made three timepieces between 1774 and 1777 based upon Harrison's invention and entered them in the Longitude Act competition. The first of these was stopped "inadvertently" by Nevil Maskelyne during testing; a month later the Astronomer Royal managed to break the mainspring. Legal action by Mudge's lawyer son, won a court judgment awarding 3,000 pounds to compensate for the damage – four years later. An improvement that Mudge devised on the Harrison originals has continued in use by the watchmaking industry until the mid-twentieth century, notably the famous Ingersoll dollar watch and later in the "Mickey Mouse" and "Timex" watches.

John Harrison died March 24, 1776, exactly 83 years after he was born. Within 10 years, "chronometers" based upon his invention were in common use with ships of commerce and recognized as far more accurate than the lunar system of measurement. When 40 years had passed, 5,000 chronometers were in use, cheap enough for a ship to carry more than one. Charles Darwin sailed on the H.M.S. "Beagle" in 1831 – with no less than 22 on board! Captain of the ship, Robert Fitzroy, a wealthy man, owned six; five he had borrowed; 11 were provided by the Admiralty.

Thus finally ends the story. H-5, the last instrument hand-crafted by John Harrison now lies in state in the Guildhall of London. Its four predecessors, H-1, H-2, H-3, and H-4 were hidden away during Maskelyne's lifetime. Neglected for nearly 200 years, they were discovered in this century, reverently cleaned and long-missing parts replaced. By February 1, 1933, all were again keeping perfect time in the Greenwich Observatory.

WERNER von HEISENBERG
Still an open question: did he
Deny the A-bomb to Hitler?

Chiswick, a quiet suburb of London where usually the loudest sound are ripples of the Thames on the river bank, rocked from a monster explosion on September 8, 1944. It broke windows for miles around, reduced ancient buildings to rubble, killed hundreds of residents. The first German V-2 rocket, launched from a mobile pad in northern France, had landed.

Three thousand clones of the 46-foot-long V-2 rocket killed thousands in England, Belgium, and liberated Paris in following months. The cigar-shaped bomb was six feet wide, filled with a mixture of liquid oxygen and alcohol. The lethal missile was guided by a set of gyros, not by radio, which could be tracked by listening devices. A 10-mile radius was then the effective zone, but accuracy would improve with experience. A separate warhead could be affixed. A nuclear warhead might then be delivered to a pinpointed bullseye. It was feared that even a few such weapons could cripple the nerve center of the British government and the Supreme Headquarters of the Allied Expeditionary Force and impose armistice upon the war-weary island.

The German rocket program had been known to the British as early as October 5, 1939. A secret inventory, listing major military science research and development programs in the Third Reich had been delivered that morning to the British Consulate in then-neutral Oslo, Norway. The report was signed by "a German scientist "who wishes you well." The identity of that scientist, privy to the innermost secrets of the Nazi machine, has never been publicized.

In the spring of 1945, Allied armies were sweeping across northern Europe, closing on Berlin from east and west. The U.S. Third Armored Division, under General Maurice Rose, raced toward Paderhorn and offshore Peenamunde. Other units headed for Dora Nordhausen, in the Harz Mountains of East Germany. These were the prime factories where rocket-scientist Werner von Braun was known to be producing the guided missiles in vast underground catacombs. Though Allied victory seemed near on the ground, the clock was ticking; fortunes of war could shift overnight. Terror weapons were Hitler's last hope, to "strike civilians day and night, break the morale of the enemy and destroy London," as he screamed.

Parts of Germany were still being stubbornly defended against Allied ground units. A team of American combat engineers, on orders from General Bedell Smith, Eisenhower's Chief of Staff, darted into enemy-held territory. They headed for Hechingan, where the mastermind of Germany's nuclear effort was – they thought – producing an A-bomb.

127

Werner von Heisenberg held the position formerly Albert Einstein's. He was director of the Kaiser Wilhelm Institute of Physics, acknowledged to be a genius, awarded the Nobel Prize for Physics at age 30. He'd been trained in Copenhagen by Niels Bohr, whose prestige in physics was only eclipsed by Einstein's.

As czar of Germany's A-bomb project, he started years early with assets far superior to America's. He had ample uranium from the world's richest mines, in Czechoslovakia. The world's only plant manufacturing heavy water was in Nazi-occupied Norway. He had authority to recruit Germany's finest physicists, chemists, engineers, and other technical talent.

All this was known in America. It would have been no surprise if Germany had detonated a uranium-fission bomb first. "We were convinced that the Nazis were far advanced in … (bomb) construction," said J. Robert Oppenheimer, head of the American nuclear program; … "possession of such a terrible weapon by that evil power would be a catastrophe for the rest of the world."

When the Americans reached Heisenberg's headquarters at Hechingen, however, they found little evidence that he had even worked on an A-bomb. They uncovered a stock of unprocessed uranium, buried in a ploughed-over field. Hidden in a grain mill, they found many drums of precious heavy-water – never used. They found underground nuclear reactors – never activated. Files of research papers were retrieved from a cesspool – but no evidence of work on advanced technology or engineering such as America's.

Heisenberg himself was missing. The task force sped to his family home in Urfeld, many miles distant. Three days later, the haggard scientist surfaced, having cycled cross-country, mostly at night, to avoid capture. Almost immediately, he was airlifted out of Germany, first to France; from there to London; and then to "Farm Hall," a secluded country mansion.

For months, Heisenberg and other top-level German scientists were sequestered. Every room was "bugged." Every word spoken, day and night, was recorded. One of the voices on the tape was that of Nobel Laureate Otto Hahn, who, in 1938, had discovered the process to split the uranium atom. Physicists everywhere learned that, by neutron-bombing of uranium, its atom would split, creating pulses of energy. Under continuous neutron bombardment, each part of the split uranium atom would split again, and again, and again, and again – each time producing energy releases, a chain reaction. Unlike fossil sources of energy – coal, oil, wood – atomic energy would be inexhaustible. Theoretically, it would produce power cheaply. Physicist Edward Teller predicted "it wouldn't pay to meter it!"

Utopia! Or? …

Ernest Rutherford, head of the Cavendish Physics Laboratory in Cambridge, England, warned, "a wave of atomic disintegration might be started … which could … make this old world vanish in smoke." Max Planck, creator of the famed physics institute in Germany bearing his name, foresaw disaster "if this instrument of power gets into the wrong hands." Facing that possible consequence of his discovery, Otto Hahn spoke of suicide.

Their apprehensions were well-founded. On April 29, 1939, <u>four months before the invasion of Poland</u>, a task-force in the German Ministry of Education resolved to "draft" Germany's leading scientists. Total mobilization of brains and resources would be applied to harness atomic energy for high-yield explosives.

German scientists, visiting at the Cavendish that last peacetime summer, described these events to their English hosts. Dr. R. S. Hutton, a metallurgist on staff, in his <u>Recollections of a Technologist</u> summarized his memories of those whispered warnings: "Hitler considered ... an atomic bomb as his secret weapon number one."

In the summer of 1939, the U. S. Army consisted of 227,000 soldiers, only one-third of them equipped —- with outmoded weapons. "A few nice boys with B-B guns," as TIME magazine reported. The United States was still struggling to come out of the Great Depression. Millions of men were still unemployed. The country overwhelmingly was "isolationist" – opposed to getting involved in the ancient animosities and ethnic rivalries of Europe. The popular entertainer, Eddie Cantor, signed off his Sunday night radio hour every week by singing "Let Them Keep It Over There." Charles Lindbergh, America's national hero, headed "America First," an isolationist organization with a huge membership.

Heisenberg visited the United States that nervous summer of 1939 to meet with old European friends. From Enrico Fermi, he learned that the U.S. Navy had rejected his proposal to commence nuclear research. Hungarian Leo Szilard bitterly admitted that most of the European émigrés had no job. (When his own part-time teaching assignment ended at Columbia, even permission to use the labs had been withdrawn.)

But Nobel Laureate Arthur Compton assured Heisenberg that, however slow off the mark, America would join Britain and France if war broke out; though it would take time for the public opinion to make a U-turn.

Heisenberg thus concluded that, even if the United States joined the war, it would not bring to Europe a first-strike nuclear capability. That being the case, what should he do, if called upon to work on atomic weaponry for his Fatherland? Should he, or should he not, help to provide Adolf Hitler with the means to enslave all of Europe? The dilemma was defined by C. P. Snow, Britain's eminent science historian: "It is not enough to say scientists have a responsibility as citizens. They have a much greater one than that, and one different in kind. For scientists have a moral imperative ... It is going to make them unpopular in their own nation-states. It may do worse than make them unpopular ..."

Heisenberg detested Hitler's ideology. Snow describes him as "coming from the elite of German academic life ... literature and music were part of the air they had always breathed." Before Hitler, his social class was devoted to ideals of academic humanism. His father, a university professor, urged him to shun contact of any kind with the Nazis. Which he had. For six years after Hitler took power in the early 30s, he had avoided collaboration. He had instead used his prestige to find employment for Jewish scientists and university professors, deprived of a livelihood. He had publicly repudiated and ridiculed nazi party-hack opportunists in the sciences who denigrated the work of Einstein and Bohr as "Jewish physics" – and thus earned curses as a renegade. The nazi party newspaper editorialized that "white Jews" like Heisenberg should be made to "disappear." As German science became increasingly politicized, he was repeatedly grilled by the Gestapo.

C. P. Snow recalled: "With discovery of fission, and with some technical breakthroughs in electronics, physicists became, almost overnight, the most important military resource a nation-state could call on." Heisenberg's reputation as an anti-Nazi was known to the émigré scientists whom he visited in America. Old friends pleaded with him: stay here! Those who had senior university posts offered him any position he wanted, at any prestigious campus – the same deal that brought Einstein to Princeton.

To each such proposal, he responded that he knew war would be a disaster for Germany. But it was his duty to remain there and create "islands of decency" (his words) until Hitler's regime fell. Then, after the debacle, pick up the pieces and rebuild. These pre-war resolutions were unanimously recalled postwar by the European-born scientists.

Heisenberg sailed back to Germany in August 1939, on the last ship to cross before the blitzkrieg against Poland. Then he disappeared for years from the "radar screen" of those tracking the sciences in Germany.

The émigré scientists in America knew that, if anybody in the world could harness atomic fission to a weapon capability, it would be their one-time friend. Would he? Could he not?

In a radio broadcast, Hitler promised soon to "employ a weapon against which no defense would avail." C. P. Snow predicted in "Discovery" magazine, September, 1939, that "within a few months, scientists will have produced for military use, an explosive a million times more violent than dynamite. It is no secret; laboratories in the United States, Germany, France, and England have been working on it feverishly since the spring."

Leo Szilard was obsessed by the nightmare. He visited fellow-Hungarian (and Nobel Laureate) Eugene Wigner, newly-appointed Chairman of Physics at Princeton. They went to see Albert Einstein together at his weekend home on Long Island, and they so alarmed him that he wrote a now-famous letter to President Roosevelt, warning that "In the course of the last four months, it has been made probable ... to set up a nuclear chain reaction in a large mass of uranium by which vast amounts of power and large quantities of new radium-like elements would be generated ... in the immediate future ... extremely powerful bombs of a new type may thus be constructed ... (that) might very well destroy (cities) ... together with some of the surrounding territory." The letter was entrusted for hand-delivery to Alexander Sachs, a friend of Roosevelt's. Though the U.S. was still officially neutral in the war, the president wasn't. Looking to the future, he appointed an Advisory Committee on Uranium. In leisurely fashion, the Committee held hearings. Nothing much was learned. Fermi and Szilard were, at first, barred from attending because they were foreigners.

Security provisions were as witless in England as in America. German refugees Rudolf Peierls and Otto Frisch, working in the Physics Department of Birmingham University, had invented the formula to produce uranium-235. But they were denied appointment to MAUD, the English equivalent of the Manhattan Project; they couldn't discuss or explain their own work to those who were working on the bomb.

From the first days of September, events were moving rapidly. Hitler's shadow spread terror across Europe. America still slumbered, but scientists were waking up. In the February, 1940, edition of Physical Review," Niels Bohr, working at Princeton with young John Archibald Wheeler (who had "interned" with him in Copenhagen) published a brief outline of the effect upon uranium of neutron bombardment. A more complete description followed in June, identifying a new element – the 93rd – created as a by-product of the collision with isotope U-238.

Next, a 94th element was developed, named plutonium. (Heisenberg's former student and colleague, Carl Friedrich von Weizsacker, applied a different name to a similar formula he created in Germany almost simultaneously.)

When a senior physicist at the Institute decided to emigrate rather than surrender his passport from his native land, he reported to Warren Weaver, of the Rockefeller Foundation, a major funding donor for the Institute, that the army's atomic research program was building "an irresistible offensive weapon." Enroute to his new job at Cornell, the informant told Szilard the same dread news. Work on the nuclear explosive device was proceeding on an accelerated basis.

Maddened by the Uranium Advisory Committee's ho-hum attitude, Szilard persuaded Einstein to send a second letter to Roosevelt which warned: "Since the outbreak of the war, research (on uranium) is being carried out in great secrecy ... at the Kaiser Wilhelm Institute of Physics ... which has been taken over by the government ..."

Einstein's second letter was dated March 7, 1940. Roosevelt issued an Executive Order that transferred responsibility for uranium research from the slow-motion civilian "Committee" to the U.S. Army.

Germany invaded Norway April 10, 1940. Major objective: seize control of the world's only purpose-built plant capable of producing "heavy water." The

New York Times explained the importance of the target: "Heavy water ... provide(s) a means of disintegrating the atom ... resulting in a powerful destructive force."

That small cluster of buildings at Rjukan, on the coast west of Oslo, was thereafter regularly pummeled by Allied air power. The Nazi High Command ordered Heisenberg: design and build our own plant in Germany. He launched a basic research project to study the absorption rate of uranium in heavy water. Patents on the process had, for years, been held by the Frenchman Frederic Juliet-Curie who was building Europe's first nuclear reactor. Whatever information Heisenberg needed about heavy water was available via a phone call to nazi-occupied Paris. This curious "oversight" could be interpreted that Heisenberg was in no hurry to produce a nuclear weapon.

Instead, he designed a reactor, immersing uranium rods in a tank of heavy water. With this design in hand, he went to see his old friend and mentor, Neils Bohr, in Nazi-occupied Copenhagen. The distinguished scientist was half-Jewish and an outspoken anti-nazi. In her book Inner Exile, Heisenberg's wife Elizabeth, says her husband "wanted to signal to Bohr that ... (he would not) build a bomb ... He hoped that the Americans ... would perhaps abandon their own development ... (and) ... that his message could prevent the use of an atomic bomb on Germany one day."

In "Heisenberg's War", Pulitzer Prize historian Thomas Powers sees another reason Heisenberg sought the meeting. He "betrayed ... (existence of) the German bomb program." Mark Walker agrees, in "Nazi Science: Myth, Truth and the German Atom Bomb."

The exact words between Bohr and Heisenberg will never be known, but America's "Manhattan Project" chiefs, General Leslie Groves and J. Robert Oppenheimer, grew increasingly worried. Their program was history's costliest investment in science – an all-out commitment to produce a nuclear weapon before the Germans did. Shortly after Heisenberg's visit to Denmark – and the U.S. entry into the war, Groves was put in supreme command of the nuclear program. From the very beginning, he was almost paranoid about the German "lead" in the A-bomb race. He was barely dissuaded from sending a fleet of bombers – at predictably great cost in airmen's lives – to destroy the Kaiser Wilhelm Institute for Physics in Berlin, where the German nuclear-energy program was then thought to be centered. (It wasn't.) He sent an OSS "hit-man" (Moe Berg, the polymath baseball pro) to assassinate Heisenberg in Switzerland. He didn't, after a private talk, which convinced him that he was not working on a bomb.

Though the German A-bomb program was standing still, nothing else was. By floating false rumors that Britain was at the point of suing for peace, the wily Churchill stalled Hitler's invasion of the Soviet Union, ready in June, 1942, until late summer. So sure had Hitler been of quick victory, that when he finally ordered the march on Moscow, his general staff had not made provision for a winter campaign. At the cost of 250,000 dead, the Red Army stopped nazi columns all along a 1,000-mile front. Within sight of the church spires of Moscow and Leningrad, the German blitzkrieg sank to a halt, axle-deep in mud, then snow.

Wehrmacht soldiers wrote home, describing their misery. No warm clothing. No winter footwear. Partisans harassed German supply lines all across Poland, Russia, and the Ukraine: food supplies ran short. Precision-machined German artillery froze in sub-zero temperatures; shells would jam or mis-fire in the barrels. Motorized equipment that had easily roared through western Europe was helpless.

The months dragged on. Millions of American men and women donned uniform. Thousands of factories were converted to war work. Three cargo ships per day were sliding down the ways in shipyards all over the country; military materiel began to reach Europe in huge volume. German submarines were being targeted by newly-invented radar and sonar, and picked off —- one a day. Hundreds of four-engine B-17 "Flying Fortress" bombers were crossing the Atlantic. Each would carry 8,000 pounds of bombs. The P-47 pursuit plane, mounting eight machine-guns, could out-maneuver Luftwaffe pilots; its firepower could neuter heavy tanks and overturn heavy trucks.

In Churchill's ringing words, it was the "end of the beginning, the beginning of the end." The Fuhrer and his generals had to face – for the first time – the likelihood of a long war. Recognizing that every possible resource had to be concentrated on the Russian front, Reichsmarshal Hermann Goering, (Hitler's #2 man), issued orders to curtail support for all industrial and technologic projects – if they could not achieve operational effectiveness within nine months.

That edict was amazingly effective. Three years later – despite day-and-night Allied bombing of every visible industrial target – Germany was producing more tanks, artillery pieces and ordnance than at the beginning. But not producing atomic bombs. Though Werner von Heisenberg was, officially, top man in the uranium program, other ministries in the German government and rival German scientists were pursuing alternate lines of uranium research – and competing for resources. Heisenberg's own inertia was cloaked by the confusion.

He issued reassuring progress reports. He was on record as asserting that a fission device was now "theoretically practical." His colleague, von Weizsacker, further said that the new element (plutonium) would build "an explosive of unimaginable power." For over two years, a stream of such optimistic statements lulled the nazi hierarchy into believing that a big bang climax to the war was in sight.

With mounting casualties reported from the Eastern Front and bomb-debris mounting high in every German city, Albert Speer, Goering's Minister of

Armaments Production, was forced to re-evaluate war-production priorities. An architect by profession, Speer had, in the thirties, delighted Hitler with the Wagnerian stage-set he'd built for Party Day rallies in Nurenberg. (Even during the war, he entertained his master with designs for magnificent new state buildings to be built post-war: new cities; face-lifts for old ones. He was, by temperament, strongly attracted to the drama of a nuclear weapon, winning the war with one mighty crescendo, like a thunderbolt from Thor, the mythological Teutonic god. He convened a conference of political and military heavyweights whose influence was necessary to override Goering's vetoes and persuade Hitler that a nuclear device would do just that. He wanted development work to be accelerated, not terminated.

In advance of a do-or-die meeting with Speer, a Heisenberg agenda was circulated. His men used technical jargon so abstruse that most of the high-level invitees didn't attend. But, at the meeting, Speer heard what he wanted to hear: "nuclear" and "practical." Having been told that, with the new element (plutonium), "the road was now "clear" to making a bomb. He summoned Heisenberg to another meeting and demanded – how quickly? What did Heisenberg need from him, Speer, to expedite a crash effort?

The question that had haunted Heisenberg since 1939, when he learned that America was not building a first-strike arsenal, now had to be faced.

There were, Heisenberg said, many, many technical problems in translating nuclear theory into nuclear weaponry – many unknowns. The world – or most of Germany – might blow up if one of the experiments went wrong. Shipment of vast quantities of materials would clog the already-overtaxed rail network. An army of scientists and technical personnel would have to be released from military service. Time would be needed. Lots and lots of time, he repeated, alluding to Goering's taboo against diversion of resources from the Eastern Front.

During a 1967 interview published in the German magazine "Der Spiegel," Speer says he "asked Heisenberg to put together a list of material and financial demands (to expedite work)." Heisenberg responded with requisitions "so ridiculously tiny ... that we got the view ... the physicists themselves didn't want to put much into it." One of Speer's assistants who was present at that fateful meeting was a chemist named Lieb. Grilled by U.S. Air Force Intelligence immediately after the war, he reflected: "the interest of Speer ... lessened (because) no way was found to expedite the development." While serving his 20-year prison term as a war criminal, Speer wrote "<u>Inside the Third Reich</u>." He tersely concurred: " ... we scuttled the project to develop an atomic bomb."

German physicist Friedrich Houtermans, when interrogated by U.S. Army Intelligence after the war, vowed that Heisenberg and Weizsacker had reassured him that his formulas for the plutonium-production process would be protected if it came to their "official attention." Heisenberg had, all along "put the war in the service of science," Houtermans insisted, under oath. The commando unit that reached Heisenberg's laboratory in Hechigen confirmed this: eight small power-generating nuclear reactors had been built. America did not build its first such energy-generator until years later.

After the Speer meeting in June, 1942, Heisenberg's routine was to visit the Physics Institute in Berlin only once a week. He never stayed more than two or three days. (In contrast, Oppenheimer worked a 20-hour day, seven-day week; and weighed barely 115 pounds by the end of the war. He could sit comfortably in a child's high chair.)

All components in America's atom-bomb Manhattan Project, scattered nationwide, were integrated in constant communication. Whereas units in Heisenberg's enterprise were autonomous. He supposedly "supervised" hundreds of physicists, chemists, engineers, mathematicians, metallurgists, and technicians – but never organized them into a cohesive team. Each workstation poked along independently at its own speed, oblivious to what others were doing – or the possible relevance thereof to their own work. (When Heisenberg's man in charge of producing U-235 isotopes asked for a small amount of uranium to proceed with his experiments, his requisition was denied by Heisenberg himself, who personally controlled the entire stockpile of uranium ore.)

Heisenberg has never claimed credit for the critical role he played in the German A-bomb drama. His postwar book, Physics and Beyond, is reticent. Were he to admit, publicly, he had deliberately steered his assignment into impotence, it would be a confession to treason. Lack of a reprisal terror weapon left the Fatherland open to colossal destruction. Every city, every rail hub, every airport, every bridge, every industrial center in Germany had been reduced to rubble. The nation itself had been dismembered. Disease and hunger were endemic. Every family suffered losses.

The German public already knew that the nation's most famous physicist had, somehow, failed to produce. They didn't know – for sure – that his failure was purposeful. Reviewing all the evidence, Jeremy J. Stone, President of the Federation of American Scientists, interprets: "German atomic scientists ... did not want (their countrymen) to hear ... evidence of their dissidence ... nationalists were ready to cry 'sold out the nation'." (This, of course, was the Nazi explanation for Germany's defeat in World War I.)

"Patriotism" has no dimension. Nor does chauvinism. America is not exempt. Reflecting upon the 55,000 Americans who died in the Vietnam War, Robert McNamara – then Secretary of Defense – now says the war was "wrong – terribly wrong." He admits he knew it at the time, but that it was his patriotic duty to be loyal to his leader. (The word "leader" in German is "Fuhrer.")

Few are those in public life strong enough to swim against the tide. In Germany, physicist Max von Laue, a Nobel Laureate, flatly refused to work on a Nazi bomb. In America, Oppenheimer's close friend, Nobel Laureate physicist,

I. I. Rabi, could have had any post in the Manhattan Project he wanted – and accepted none. The purity of their moral stature is obvious and admirable.

Oppenheimer and his team of scientists – many of them refugees from Hitlerian aggression – had good reason to hate Germany, and fear the consequences to all mankind if the dictator had prevailed. Heisenberg and his close associates had no such motivation. Though they were anti-Nazis, they were trying to survive in a fascist society until the regime fell. We will never know whether they could have produced an A-bomb – if they had wanted to. Those who think yes – as this writer does – can point to the advantages Heisenberg enjoyed: years of lead-time from Otto Hahn's 1938 atom-split discovery; ample supplies of uranium and heavy water; early possession of the plutonium secret. Most important: the Nazi dictatorship could mobilize whatever manpower and facilities Heisenberg wanted as was done for von Braun's guided missile project (32,000 slave laborers built the underground facilities in eight months!)

This is theoretical. The Manhattan Project in the U.S. required vast investments of equipment, facilities, men, and money – in a country with intact communications and transportation systems. 500,000 people were employed directly or indirectly. Very different was the situation in Germany after American air-might joined the R.A.F. in 1942. So it is uncertain whether Heisenberg could have built a bomb before Oppenheimer – even if he'd wanted to. This question still provokes heated debate among scientists. At one extreme, American physicist Samuel Goudsmit, whose Dutch parents died in a concentration camp, studied the German equipment and lab records. He concluded that Heisenberg's team simply did not understand the physics, chemistry, and engineering necessary for conversion of a chain reaction into a nuclear explosive. Whereas Hofstra University professor David Cassidy, in "The Life and Sciences of Werner von Heisenberg," draws upon documents long buried in Rockefeller Foundation archives and concludes that Heisenberg's cadre pursued a calculated policy of frustrating the military's demands. In the "Journal of the Federation of American Scientists (Nov/Dec 1994)," he reminds us that "Heisenberg was a normal, cultured individual who was caught up in ... a dreadful situation ..."

In 1960, addressing the American Association for the Advancement of Science, C. P. Snow summed up: "I admire in scientists very simple virtues – like courage, truth-telling, kindness ... Whether they like it or not, what they do is of critical importance for the human race ... They may not have asked for it, or only have asked for it in part, but they cannot escape it. They think, many of the more sensitive of them, that they don't deserve to have this weight of responsibility heaved upon them. All they want to do is get on with their work ... but (they) can't escape the responsibility ... the gravity of the moment in which we stand."

In a dialogue analyzing "virtue," Socrates defined courage as a quality beyond fearlessness or bravado. True courage will face – and feel – fear; face the consequences of action; and in full knowledge of risk and cost, take action.

Or not take action. It is an ancient and honorable moral policy, as enunciated more than 2,000 years ago:

> Do not unto others
> What you would not have
> Others do unto you.
> - Hillel the Elder

SIR WILLIAM LAWRENCE
Paved the way for
Charles Darwin

Many doctors in the audience at the Royal College of Surgeons that day in 1816 had been reared in rural England. Tending sheep was one of the usual farmyard chores assigned to boys. Hence, they were amazed to hear that farmers in far-off New England had developed sheep with such short legs that the animals could be confined by low meadow walls. They had trouble believing that smallholder rustics across the ocean had bred sheep – the lecturer said – that had such "singular proportions and appearance ... (so as) in succeeding generations – their offspring had, in many instances, the same ... shortness of the limbs and length of the body ... in consequence, the animals were less able to jump over fences."

Surprises continued: "birds swallow stones to grind their food" ... and trout, too, swallowed stones to smash shellfish in order to more easily digest them.

Whimsical observations like these studded William Lawrence's series of lectures in 1816 and contributed to his immense popularity as a speaker. At 32, he was a juvenile among seniors –– but already a full professor. Doctors decades older listened respectfully. His advancement had been meteoric.

His father was Chief Surgeon in the city of Cirencester. He arranged for his boy to be indentured to John Abernathy, famed for his medical knowledge and his irascible nature. Young William soaked up the former, and survived the latter sufficiently well to be endorsed by his mentor as a "demonstrator" in anatomy at London's St. Bartholomew's Hospital. Apparently his "demonstrations" were laced with valuable information: he was invited to the staffs of the Royal Hospitals of Bridewell and Bethlehem. The Royal College of Surgeons nominated him as professor of anatomy and surgery. He was on par with Abernathy, who had been named to the same eminence earlier.

On this whirlwind of whiz-kid advancement, he inaugurated his career as a lecturer. Medical men rarely are gifted as public speakers. Lawrence was. When he commenced lecturing on anatomy at the Royal College, his unconventional style and youthful enthusiasm quickly won a following. A medical eminence, Sir James Paget, described the Lawrence presentations as "admirable in their well-collected knowledge, and even more admirable in their order, their perfect clearness of language, and the quiet attractive manner in which they were delivered ... It was the best method of scientific speaking that I have ever heard."

Lawrence insisted that every doctor – not just surgeons – should pursue "zoological study, the treatment of man as an animal, as an object of natural history ..." because this method "is the only proper foundation for teaching and research in medicine, in morals, or even politics." This credo was seen by medical men as an attack on religion by those who feared the light of science on church doctrines rooted in the Bible. It was seen as subversive to the monarchy and aristocracy.

"We cannot expect to discover the true relations of things," Lawrence said, "until we rise high enough to survey the whole field of science, to observe the connections of the various parts and their mutual influence." Dangerous talk – as heard by those with vested interests in status quo. What was not wanted was elevation of the common man to a height where he could see the connections between landowners, church and state, and recognize that all life in Britain was dictated by that oligarchy.

Lawrence asserted that all mental activity was as much a function of the brain as digestion is of the stomach. Human growth follows the same laws as the plant world: "The ascending sap ... (is) equally fitted to form the leaf, the flower and the fruit ... the genial warmth of spring ... cause(s) the ascent of the sap, the plant seems bursting with life ..." This line of thought was thought to be "materialism," the spawn of French philosophy.

Lawrence's lectures were collected and printed in 1816 as a book: "Introduction to Comparative Anatomy and Physiology." It attracted wide readership – much wider than just physicians. Reception was stormy. Some found it offensive that Lawrence viewed the human brain as functioning like any visceral organ. A howl greeted his suggestion that man is fundamentally an orang-utan with somewhat more "ample cerebral hemispheres." These were "disgusting caricatures" of the human species; critics rejected Lawrence's demotion of the "highest and noblest properties of man's nature ... (to a level) with the anthropoid animals ... Horses and asses, oxen and sheep, dogs and hogs, rabbits and poultry."

Dr. Abernathy used his superior post to deride Lawrence's zoological arguments, so different from those of inherited tradition.

Church of England clergy were nervous about teaching that relied on observation, logic and reason. The French Revolution had unchained men's minds from ancient bonds, including the folklore and mysteries of religion. Anti-clericism was widespread, an "infection" that, it was feared, would affect more people as teachings such as Lawrence's became popular.

The popular young doctor overnight became a leper. From hundreds of pulpits, and in the literary press, Lawrence was denounced for denying the existence of the human soul; for flying in the face of Biblical Scriptures; for being anti-clerical. He was accused of being a "Free-thinker." Of being pro-Quaker. Of secretly favoring the American system of government. It was suspected that he was a disciple of the half-Jewish Frenchman, Montaigne, who had advanced the idea that man is in no way superior to brutes emotionally or intellectually; and he mocked the Cross and Original Sin and the Flood as but "empty shadows of ... religion."

Christianity was – and is – the state religion in England. Lawrence's teachings were, as often before in history, a critical dialogue between reason and faith. (Some churches at that time refused to install Ben Franklin's lightning rod, rather than interfere with God's will.)

William Lawrence was relieved as Lecturer at the RCS. Dr. Abernathy took over, and continued to assail his former protégé's radical ideas. But compared to the charismatic Lawrence, he was a dud. Audiences shrank drastically. Lawrence was invited to resume lecturing – but was enjoined to continue with lectures confined to traditional knowledge. He agreed – but first, he had a score to settle:

> "To fair argument and free discussion, I shall never object, even if they should completely destroy my own opinions; for my object is truth, not victory. But when argument is abandoned and its place supplied by an inquiry into motives, designs and tendencies, the case is altered ...
>
> What are the overt acts to prove ... my treason against society ... What support can you discover for ... imputations (against my) profession, pursuits, habits and character? How will it promote (my) interests to endanger the very frame of society? By what latitude and artifice of construction, by what ingenuity of expression can the materials of such a charge be extracted from the discussion of an abstract physiological question?"

Then Lawrence leveled heavy artillery on those whom he identified as enemies of scientific enquiry: "I will not be set down nor cried down by any person, in any place, or under any pretext ... if it involves any sacrifice of independence, the smallest dereliction of the right to examine freely the subjects on which I address you, and express fearlessly the result of my investigations ...

"Instead of wishing or expecting that uniformity of opinion should be established, I am convinced that it is neither practical nor desirable; that varieties of thought are as numerous and as strongly marked, and as irreducible to one standard, as those of bodily form; and that to quarrel with one who thinks differently from ourselves would be no less unreasonable than to be angry with him for having features unlike our own ...

"French physiologists ... seem to be considered our natural enemies in science, as well as in politics ... unfounded charges and angry invective, undisguised and glaring national partiality, unreasonable national antipathy, unmerited and unprovoked abuse of the writers of a whole nation, afford an overwhelming proof of its complete moral inefficacy."

Lawrence's vigorous counter-attack reminded his listeners that "the greatest of the ancient philosophers said that the surest way of gaining admission into the temple of wisdom as through the portal of doubts; and he declared that he knew only one thing – his own ignorance ... (Whereas) in modern philosophy, doubting is proscribed, as the source of all mischief; and an overbearing dogmatism, even on the most abstruse and difficult questions, is held forth as a wiser course than the modest confession of ignorance ...

"By a curious inconsistency in the human mind, difference of opinion is more offensive and intolerable in proportion as the subject is ... less susceptible of direct proof. Hence the rancorous intolerance excited by ... shades of opinion that distinguish many religious sects ... The Priests ... rain a torrent of abusive epithets, as heretic, infidel, atheist, and the Lord knows what, on all who (have) audacity to differ from them in opinion. This ecclesiastical artillery has been so much used, as to have become, in a great measure, unserviceable; ...

"I take the opportunity of protesting, in the strongest terms – in behalf of the interests of science, and of that free discussion which is essential to its successful cultivation – against the attempt to stifle impartial inquiry by an outcry of pernicious tendency – and against perverting science and literature, which naturally tend to bring mankind acquainted with each other, to the anti-social purpose of inflaming and prolonging national prejudice and animosity ...

"Science, the partisan of no country, but the beneficent patroness of all, has liberally opened a temple where all may meet. She never inquires about the country or sect of those who seek admission; — she never allots a higher or a lower place, from exaggerated national claims or unfounded national antipathies. Her influence on the mind like that of the sun on the chilled earth, has long been preparing it for higher cultivation and further improvement."

It was not the first time, it will not be the last, that an open mind attempts to debate with closed minds. Lawrence had the youth, the self-confidence and courage not only to defend himself, but to go on the attack. He continued his lectures in that spirit. Until ...

In 1818, he invaded a "no-go zone" – assailing the snobbery and class-consciousness of British society. His new series of addresses at the Royal Society were published as "Lectures on Physiology, Zoology and the Natural History of Man."

It was widely believed in those days that high-caste individuals were larger in size, stronger, and better-looking because of divine selection. "Well-born" men and women benefited from adequate nutrition, adequate shelter, adequate clothing, plenty of rest, fresh air and outdoor exercise. More of their progeny thus survived the diseases of infancy. Few died in childhood from malnutrition or exposure. Wealthy young men were in a position to pick their brides for looks, titles, bloodlines, health, and money. With generations of selective breeding, many in the "upper class" looked to be a breed apart.

But Lawrence drew attention to the downside. The mating habits of Europe's aristocrats were causing, he said, a "degradation of race" in princely families. He reminded readers that "the destiny of nations are entrusted" to them; on their qualities and actions depend the future happiness of millions. "Law, customs, prejudices, pride (and) bigotry confine them to intermarriages with each other ... and ... all the pernicious influences inseparable from such exalted stations ... The strongest illustration of these principles will be found in the present state of many royal houses in Europe."

The king, George IV, was an obese gargoyle, of whose death in 1819 the London Times commented "there never was (loss of) an individual less regretted by his fellow creatures ..." His brother William's illegitimate children all died as infants – but he had 10 "legitimate" ones. Another brother, Ernest, the Duke of Cumberland, was suspected of having raped his sister and – it was assumed – murdered his valet, witness to the prank. The predecessor of this motley crew was George III, certifiably insane, abhorred by American colonists, but very popular with the ladies as were his sons. Fifty-six of the 59 children sired by the royal stud were illegitimate.

Lawrence was not being judgmental; in his lectures and their printed texts, he stressed that man was fallible – a part of nature, subject to analysis and criticism as any other component of natural history. "Food, climate, mode of life, have the power of modifying the animal's organization." Long before he published "Origin of the Species," – decades later – Darwin credited Lawrence as having established this principle. In Darwin's subsequent book, "The Descent of Man," Lawrence is further cited as having originated the principle whereby the comely appearance "of the upper classes in England is due to the men having long selected the more beautiful women."

But in 1818, Lawrence was "premature" by a half-century. He was carrying bold thought and blunt speech too far, too fast. His disrespect for "traditional" beliefs was assailed by an Oxford don: "How far however we can really benefit mankind by convincing them that the soul is mere perishable matter, like the body, may appear problematical to a large portion of society, for who does not know that we must attribute the French Revolution, with all its horrible attendants of anarchy, despotism and murder, to the persuasion that there was no future existence?" The polemic linked, in its title, "Surgeon Lawrence" with "Infidel (Thomas) Paine" and "Lord Byron." The ruling classes landed on Lawrence with hob-nailed boots. "The present volume had not been in circulation a month," The British Critic reported, "before it was withdrawn and the remaining copies bought up at a considerable expense, under a positive engagement ... (promise) either to alter certain passages in the work, or to withdraw it altogether from the public. Lawrence had placed himself beyond the protection of the law because he appeared to contradict the Scriptures."

Lawrence's book was banned as blasphemous. His copyright was withdrawn. He was threatened with revocation of his major hospital affiliations – unless he retracted and apologized. Lawrence also had been on staff at a mental hospital which suspended him because "orthodoxy of opinion was essential to practical skill in the cure of disorders ... under their charge."

Still young, Lawrence's career was on the brink of being aborted. Under pressure from enemies and cajoled by friends, he caved it. He confessed that "experience and reflection had only tended to convince him more strongly that the publication of certain passages in the lectures was highly improper." He withdrew his remarks; promised never to reprint them; never to publish anything more of similar nature.

Though "highly improper" ideas in scientific texts would not be cause for draconian punishment today, it appeased the "Establishment" then. William Lawrence's cautious retreat dampened the howls that could have inflicted worse damage.

For months, for years, witnesses commented in professional journals about the contretemps, a foretaste of the storm that would break over Darwin's head 40 years later. The progressive movement in science emerged in defense of Lawrence in professional journals, but...

" ... The governors of the institution ... having decided that Mr. Lawrence's opinions are of a dangerous tendency, suspended him for two of his appointments, and there is no saying how much further they would have proceeded, had he not appeased them by suppressing his book ... he stated that he acted in deference to the opinion of his friends, who considered his work as having a bad religious tendency ... With this partial compromise, the disputes ... subsided ... The facts of science must be tried by their own merits, by their consonance to nature, which is always an infallible guide."

" ... it is the province of reason to set us right ... any other authority in matters of science is always dangerous ... It may restrain the actions, but cannot fetter the mind. It may impede the march of knowledge, but cannot extinguish it ... the less the mind is inured to reasoning, the more easily will it be operated upon by the passion of fear."

"The subject in dispute was ... nothing less than the origin of the vital principle in man, or the immediate cause of the phenomena of life ... not only has a stop been put to inquiry, but religion has been brought in to inflame the passions, and confirm the prejudices of another generation."

Abuse of the chastened doctor continued for years after his "retraction." He was derided by conventional thinkers: "none but a philosopher mistakes a man, however humble his intellect, for an improved variety of monkey." As late as 1850 – 30 years later! – the press was still flogging the animal analogy: "we confidently predict that ten years will not have elapsed before it will be regarded with ... contempt."

He never answered back. Instead, he confined himself to non-controversial science. In May 1824, he was appointed surgeon to St. Bartholomew's Hospital – an office he held for over 40 years. From that prestigious post, he cultivated a wealthy clientele. When the ophthalmoscope was invented in 1851 in Germany, he widened his areas of practice to include eye diseases and surgery. He was elected more than once to the presidency of the Royal College of Surgeons. He was the first to be appointed the queen's personal physician, and be knighted.

Alfred Russel Wallace is credited as being co-originator, with Charles Darwin, of the theory of natural selection (evolution). In a letter dated December 28, 1845 – more than a generation after Lawrence fell silent and before "Origin of the Species" was published – Wallace observed "that many eminent writers give great support to the theory of the progressive development of species in animals and plants. There is a very interesting and philosophical work bearing directly on the subject – Lawrence's "Lectures on Man" ... which are now published in cheap form." (The banned book!) In a letter to a friend, Wallace recognized the great principle: variation among individual creatures is prelude to variation within a species. Ergo, evolution. He concludes his letter: "Read Lawrence's work – it is well worth it." Lawrence's illegally printed book was popular with progressive readers for years after the Darwin book was available; it was last reprinted in 1866. In the final years of his life – he lived until 1867 - Lawrence was silent during the great 10-year debate that followed publication of "Origin of the Species." But not forgotten. He was quoted by Sir Charles Lyell and Sir Joseph Hooker, world-famous scientists and Darwin's closest friends, who had arranged the publication of "Origin." Another Darwin intimate, Thomas Henry Huxley, in the introduction to his own book, "Man's Place in Nature," paid tribute to

> "Sir William Lawrence ... one of the ablest men who I have known ... well-nigh ostracized for his book "On Man," which now might be read in a Sunday-school without surprising anybody."

A portrait of William Lawrence hangs at St. Bartholomew's Hospital in London; a bust is in the College of Surgeons where a scholarship and gold medal are awarded annually in his memory. But his name is missing from most histories of medicine and science. His book is now forgotten, except in the British Museum and the underground stacks of the Bodleian Library at Oxford, where this essay was researched.

LISE MEITNER
Emerges from shadow
Of the A-bomb

Soon after World War II, the Deutsches Museum in Munich mounted a photographic exhibit featuring the laboratory in which nuclear fission had been discovered in 1938 at the Kaiser Wilhelm Gesellschaft for Physics in Berlin. Centerpiece of the display was a picture of the workbench at which the epochal work was done. Under the picture, a sign read "Arbeitstitch von Otto Hahn" (Workbench of Otto Hahn). However...

It was not the workbench of Otto Hahn, but of Lise Meitner. It was not Otto Hahn who deciphered the physics of nuclear fission, but Lise Meitner. Partners for 30 years, their co-authored papers charted the course that ultimately led to splitting the atom. Their longtime assistant, Fritz Strassmann, testified that Meitner was team-leader. Yet Hahn had insisted upon solo recognition and he alone was honored by the Nobel Prize.

Research in Wannsee, outside Berlin, Germany at the Hahn-Meitner Institute, provided the writer with the history that preceded the mystery.

Lise Meitner was born in Vienna, 1878. Her formal schooling in the Austrian state system ended when she was 14. Like all girls, however bright, no future was envisioned other than that of a housewife. But the winds of change gusted open university doors to females in 1897. At age 23 in 1901, she commenced college-level study of electricity and magnetism, elasticity, hydrodynamics, optics, acoustics, mathematical physics, thermodynamics, analytical mechanics, kinetic theory of gases. She was granted a doctorate in physics in 1906, the first female in Austrian history to achieve that rank in physics.

There was no hope of appointment to a career position, but that was no deterrence. With her parents' support, she went to Berlin and appealed to the famed Max Planck for permission to audit his advanced physics classes. He didn't believe in higher education for women. Nature, he wrote, had

> "designed for woman her vocation as mother and housewife
> and ... under no circumstances can natural laws be ignored
> without grave danger ... especially in the next generation.

But Planck recognized in this shy individual a rare quality: thirst for knowledge when there was no career advantage in its acquisition. How is it possible to say no to such a person? Planck granted permission for Lise to attend his lectures. When it came time for her to move on from theoretical studies to the laboratory, he recommended her to the professor of experimental physics, Emil Fischer. In his class, she met a young chemist, Otto Hahn, who already had an impressive lab record. He had worked in England with Sir William Ramsey, discoverer of argon, and then for a year with the renowned Ernest Rutherford in Montreal. Having specialized in organic chemistry, Hahn needed help in physics to pursue his interest in radioactivity. Meitner agreed to join him. The field was wide open. The two young scientists had complementary strengths: Hahn was interested in discovering new elements and in studying their chemical properties; Meitner was absorbed in the physics of radiation.

Their first year together saw their maiden paper in a major physics journal; the second year, their second; then another two trailblazers that same year; six the next year. In five years, nearly two dozen papers headed by both their signatures circulated internationally. The freshening flow of their discoveries swelled Germany's reputation as the world center for advanced knowledge in radiochemistry and physics. They became close friends.

From 1912 to 1915, Lise was also an assistant to Max Planck at the University of Berlin. She was paid a small fee for grading students' papers and organizing seminars. When Plank's Institute attracted financing from industry for research projects in chemistry, a laboratory was equipped and Hahn was appointed its director at a good salary. Lise – a full partner in the workload – continued in a twilight zone with no job title, no pay. Her indulgent parents continued to send a small allowance year after year. She lived in a boarding house, spent little on herself, subsisted on bread and coffee. When Planck became uncomfortable with this injustice, he wangled a small salary for her – a very small salary. (Not till 1913 – six years at the Institute! – was she awarded a title parallel to Hahn's. Not with equal pay, of course.)

By now, almost a member of the Planck family, Lise met Albert Einstein in their home; he joined Planck, an excellent pianist, and Josef Joachim – a famed violinist – in performing chamber-music trios. Lise taught the group Brahms songs. It was a golden time, so sweet it couldn't last. It didn't.

When the Hapsburg Emperor Franz Josef of Austria-Hungary fumbled his way into World War I in 1914, Lise was visiting her family in Vienna. Their home was directly across from the main rail station. She watched thousands of beautiful young men in splendid new uniforms singing as they departed for the battlefields, certain they would be home for Christmas. Many were – in boxes. Trench warfare would, in four years, bleed Europe white.

Austria's ally, Germany, mobilized Otto Hahn's generation, convinced their Fatherland was under attack by the competing colonial powers of Belgium, France, and England. Hahn joined a specialist unit to develop poison gas weaponry.

Lise Meitner also believed that her side was fighting a just war. When she heard that Marie Curie in France had organized mobile X-ray units to operate in the battle zone, Lise volunteered as an x-ray nurse in the Austrian Army. But whereas Curie, in converted motor vehicles, rattled along dirt roads behind short front lines, Meitner moved from one field hospital to another in Italy and the Balkans, Poland and Russia. Worn out, she returned to Berlin in 1916.

There, a war was also being waged – between conflicting ambitions. Senior eminences in the Kaiser-Wilhelm Institute had divided into two factions. Patriotic militants wanted to concentrate on projects directly useful in the war effort – like aerial photography and advanced gas weapons. The other group, led by Planck and Fischer, fought to protect their programs, personnel, and laboratories for basic-science research. Heavy funding for the Institute came from industry; aerial photography and poison gas had limited commercial future post-war, whereas basic research in chemistry had unbounded commercial horizons.

Fischer – who only a few years earlier forbade Meitner from entering a laboratory where men were working – assigned to Lise her own radiophysics laboratory with a director's title, and a salary at parity with Hahn.

She rented an apartment.

Hahn served in the army the full four years of World War I. When he came to Berlin on leave, he worked with Meitner. They jointly announced in 1918 the discovery of a new radioactive element, the 91st, which they named protactinium. The publication carried both their names, as usual. Note: Otto Hahn was credited as co-discoverer of protactinium even though he was in the army and away from Berlin most of the time. No such courtesy was extended when, later, Meitner had to leave the work she had initiated in Berlin.

Germany was by then suffering shortages in everything – men, equipment, clothing, food. But basic research in chemistry continued without interruption. For those in the Planck circle, this was adequate nourishment. Social life flourished in private homes. (Lise was amazed at Einstein's "naïve and really quite peculiar comments on political and military prospects.")

The nation was weary. Millions of men and boys had been destroyed, but the generals fought on. The allies and German infantry had been locked in bitter hand-to-hand fighting for years in a strip of blood-soaked Eastern France. With entry of America in the war in 1917, Germany was soon forced to accept surrender. November 11, 1918, the guns fell silent.

Lise Meitner, who once was not allowed to use the same bathroom as the men did, was head of the department of radiation and physics, a regular participant in weekly colloquia with Einstein and half-dozen other Nobel Laureates. The field was exploding. After protactinium, Hahn and Meitner dissolved their partnership. Subsidized by I. G. Farben, Meitner developed a research lab as famous as the Curie Institute in Paris and the Cavendish in Cambridge, England. She was nominated for a Nobel Prize, year after year.

Enrico Fermi in Italy started to experiment with bombardment of uranium with neutrons. Hahn and Meitner, resuming their partnership in 1934, followed his lead. But as with Fermi – and Eve Curie in Paris on the same trail – their findings were inexplicable. The inner workings of the atom were not understood by the keenest minds in the worlds of chemistry and physics. (All three were splitting the uranium atom for four years – and didn't realize it! Not until Hahn was unable to understand the results of his experiments, and Meitner was able to puzzle out the physics of those results, was the atom-splitting process understood.)

The Berliners were so absorbed in their work they were nearly oblivious to turmoil in the political life of their country. Saddled with an enormous reparations debt, Germany was gripped by runaway inflation, the economy collapsed. Factories closed. Millions were unemployed. The republic hadn't the means or time to replace a discredited, but centuries-old monarchy with a democratic system. Instead, constant strife erupted between right-wing nationalists and left-wingers, both sides blaming the other for hard times.

In this chaos, an ever-increasing number of voters was attracted by the idea of a strong leader who would restore an orderly social system. Germany's bankers and industrialists wanted civil stability, too. Adolf Hitler, once jailed as a rabble-rouser, polled more votes in every election and his once-tiny National Socialist Party became ever stronger in the nation's parliament. It never achieved a majority, but with support from farther-right factions, the little demagogue with the funny mustache was appointed as chancellor January 30, 1933.

Hitler's avowed aim, spelled out in "Mein Kampf," the book he wrote years before in prison, was to purge Jews from public life. His brown shirts soon controlled the streets, invaded private homes, offices, even courtrooms, where they abducted Jewish judges and lawyers. April 7, 1933, a "Law for the Restoration of the Professional Civil Service," prohibited citizens who had as much as one Jewish grandparent, from employment in any institution receiving government support.

Starting in May, Jewish professors, Jewish scientists, artists, architects, musicians, poets, writers, were "phased out." Many non-Jews who may not previously have had strong political affiliations, adopted anti-Semitism as a means to advance their careers.

Nobel Laureate Werner von Heisenberg's consolation letter to a fellow Nobel Laureate, Max Born, when the latter was fired as director of the world-famous Gottingen Institute for Theoretical Physics, offered regrets, but reminded the now-unemployed Born that "… only the very least are affected … by the law … Certainly in the course of time the splendid things will separate from the hateful."

Though of Jewish descent, Meitner had been baptized when young; she didn't know anything about Judaism. Which didn't stop anybody from thinking of her as a Jewess. Meitner was ordered not to attend the weekly colloquium she initiated. When Hahn was invited to speak about their partnership uranium experiments at another institution, he avoided mentioning her name. When Hahn was appointed to the directorship of the Institute for Physical Chemistry, he fired all the Jews. Though he referred later to this episode as an "unpleasant and thankless task," it was a means of retaining influence where it mattered most to him. He was not a Nazi, never belonged to the party. As he later told his grandson, Dietrich, (quoted in Dietrich's biography of Hahn):

> "there was much that I disagreed with... the presence of L. Meitner did not make the situation better ... Thus at the yearly meetings (of the Institute) I was always seated in a less prestigious place than was appropriate for my position and my age and my length of service ..."

Because of her Austrian citizenship, Meitner was insulated from the Nazi "cleansing" until March 12, 1938. On that day, Adolf Hitler, born in Austria, sent German troops across the border into his fatherland, declaring that the two countries were now one. The Austrian Army didn't resist; the Austrian people were overjoyed to become part of Hitler's seemingly-irresistible cavalcade. Long a stronghold of anti-Semitism, Vienna attacked its Jews; their homes were violated, their possessions stolen. On the streets, Jews were beaten while police watched. Many Jews committed suicide.

Later that same year, Hitler forced his will upon Czechoslovakia – with the complicity of England and France – and merged its Sudenland province into Germany. Humiliated by defeat in World War I with loss of territory, the land-mass of Germany had now been greatly aggrandized – without a war. The effect was electric on those scientists who had risen through political opportunism. They could further advance by getting rid of Meitner, who was now in a precarious position. She was no longer protected by an Austrian passport. But she had many friends in high places, who respected the substantial contributions she had made to German science over the years. Carl Bosch, head of the I. G. Farbenindustrie, a major sponsor of the Institute (and later its director) urged her "not to entertain thoughts of resigning ... but to continue your very meaningful work ..." Meant well, his support served the purpose of keeping Lise Meitner in Berlin, when many left legally.

Hahn went to see the treasurer of the Chemistry Institute to guarantee her status. He was turned down. Meitner's diary entry that night was terse: "Hahn says I should not come to the Institute anymore ... He has, in essence, thrown me out."

Like a frog in a snake's cage, she now struggled to acquire a permit for legal emigration. Too late to get out. Too late to get in elsewhere. Refugee status in sanctuary countries was available only to scientists who had guaranteed university positions awaiting them. But every post, even lowly ones, had been filled.

When Planck retired in 1937, he had been replaced by Carl Bosch as president of the Institute. Bosch had long admired Meitner; he rang every bell, pulled every string, right up to the Reich Minister of the Interior, Wilhelm Frick. The answer was immediate and blunt:

> "political considerations are in effect that prevent the issuance of passport ... to travel abroad. It is considered undesirable that well-known Jews leave Germany to travel abroad ... surely (you) can find a way for Frau Prof Meitner to remain in Germany after she resigns ... she can work privately in the interests of the (Institute). This statement represents in particular the view of the Reichsfuhrer-SS and Chief of the German Police in the Reichministry of the Interior."

That person was Heinrich Himmler, high in the Hitler inner-circle. Meitner was now a marked woman, a "political."

Illegal escape was her only hope. In 1921, Meitner had been a visiting professor at the university in Lund, Sweden, under Professor Manne Siegbahn, a Nobel Laureate. No research in radioactivity had by then commenced in Sweden, and Lise's courses – from theory to lab-practice – were eye-openers. Lise became friendly with a Dutch scientist, Dirk Coster, and his wife, Miep, a graduate student. The couple had cherished that friendship all the intervening years. Coster – handicapped by cerebral palsy – traveled to Berlin to assist with her flight. Meitner packed a few summer clothes in a small bag. Hahn drove her to the rail station the next morning. Coster was there; they met as though by coincidence.

Dutch officials had been "seen to," border police made no fuss about an expired passport and lack of a visa. By 6 p.m., the evening of July 13, 1939, Lise was in a "safe house" in Gronigen, Holland. She then stayed with the Niels Bohr family in Copenhagen, Denmark, while her famous host, a Nobel Laureate, pressured Swedish Professor Siegbahn through the Nobel Foundation to arrange for her legal entry into Sweden. She was fifty-nine years old.

It finally came through. Sweden was a fraction the size of Germany as regards resources for physics research. Meitner understood that —intellectually, but not emotionally. From almost her earliest days in Berlin, she had pursued her interests independently; been given all the equipment and materials needed. Professor Siegbahn, however, had his own agenda and maintained tight control over his limited resources. She was given a low-level job at the Nobel Institute, with a small stipend. She had no equipment, no instruments, materials, collaborators, assistants, technical support. In a letter to Hahn, she reflected:

"I often appear to myself as a wind-up doll who does certain
things automatically, with a friendly smile, but with no real life
in her. From that, you can tell how useful my work is."

Meitner and Hahn wrote to each other every other day; their letters were delivered overnight. Hahn continued the Fermi-type experiments that he and Meitner had planned, bombarding uranium. Strong in chemistry, but weak in physics, his letters to Meitner solicited answers to puzzling questions. She revised and amplified his papers before he submitted them to German professional journals – without her name as co-author, for political reasons. Their assistant since 1935, Fritz Strassmann, averred later that, despite the geographical separation, "her thought processes were present even when she was not, kept alive by the criticisms, questions, and suggestions in her letters." As Strassman said, "she was the intellectual leader of our team."

Nevertheless, Hahn floundered. In October 1938, he reached an impasse. He didn't understand the major result of his latest experiment. Hahn described failure in the leading professional journal of their discipline, "Die Naturwissenschaften," of the experiment he and Strassmann had repeatedly conducted: "… we cannot take this leap which is contrary to all experience of nuclear physics."

As described by science historian Stanley Goldberg in an essay entitled "With Friends Like These …" published by the Educational Foundation for Nuclear Science in 1996, "one of the results of (his) bombarding uranium with slow neutrons was the apparent creation of … an isotope of radium." Says Goldberg, Meitner responded that he was incorrect, it was not possible … she urged Hahn to redo the experiment.

Several weeks later, a chastened Hahn informed Meitner that the earlier conclusion was apparently in error … they had created nothing more or less than a radioactive isotope of barium … Hahn pleaded with Meitner for help in understanding these results.

Hahn again wrote to Meitner October 25: "A great pity that you are not here with us to clear up (the problem.)" On November 13, Meitner traveled from Sweden to meet Hahn in Copenhagen. She refused to believe the results of Hahn's experiment and urged him to check it again. Obediently, he and the third member of the team, Strassmann, successively transformed barium chloride into five other barium salts; in not one of the experiments could the salts be separated from the barium. Radioactivity, it was concluded, produced an isotope of barium, evidence that uranium atoms could be split – which Hahn simply could not believe.

In a letter dated December 19, he wrote: "We know ... that ... the uranium atom can't actually burst apart ..." December 21 he wrote again: "... you will do a good deed if you can find a way out of this."

Meitner's nephew, Otto Frisch, a member of Niels Bohr's group in Copenhagen, went to Stockholm to be with Lise over Christmas. They took a long walk in snowy woods outside Stockholm. Frisch's recollection: "... we both sat down on a tree trunk ... and started to calculate on scraps of paper ... the (positive) charge of the uranium nucleus, we found was indeed large enough to overcome the effect of the surface tension almost completely; so the uranium nucleus might indeed resemble a ... drop, ready to divide itself at the slightest provocation, such as the impact of a single neutron." He wrote later:

"After separation, the two drops would be driven apart by their mutual electric repulsion and would acquire high speed and hence a very large energy ... Fortunately, Lise Meitner remembered the empirical formula for computing the masses of nuclei and worked out that the two nuclei formed by the division of a uranium nucleus together would be lighter than the original uranium nucleus by about one-fifth the mass of proton ... whenever mass disappears, energy is created ... so here was the source for that energy; it all fitted!"

When they arose – in about an hour, she later estimated – they had theorized that the atom had split into two sections, roughly equal in size. Hence the evidence of barium, otherwise not possible.

All along, Hahn, as he later admitted in writing – had been calculating incorrectly: "Not being physicists, we thought of uranium's atomic weight (238) rather than the number of its protons (92)." His fixation on atomic mass had led him astray. Not Hahn or Strassmann, not even Fermi, had gone the extra step to learn that the very tiny amount of uranium used in their experiments had split, releasing violence. Back in Copenhagen, Frisch devised experiments further to prove the hypothesis.

Having verified the theory of splitting the atom in Bohr's laboratory, Frisch wrote the explanation to Hahn. The man who hadn't previously understood the physics of his experiments now became able to explain them, correctly, and rushed into print in Germany. Meitner and Frisch wrote two brief letters to "Nature" magazine, which were published with the date of February 11, 1939, to explain the Hahn-Strassmann experiments. The first was entitled "Disintegration of Uranium by Neutrons: A New Type of Nuclear Reaction." Frisch alone sent a second note, describing the experiments in Copenhagen, entitled "Physical Bombardment." It was dated February 18, 1939, a week later. The two letters confirmed that radioactivity was produced by the Berlin team from splitting of uranium into lighter elements. Meitner and Frisch, with these two articles, invented the term "fission" to describe the reaction from neutron irradiation.

Knowledgeable about nuclear structure – Hahn wasn't – and intimately familiar with masses, energy, and atomic numbers, Meitner's calculations determined that splitting the uranium nucleus had released 200 million electron volts – 20 million times more than TNT of equal amount. For the first time in history, more energy was released by the nuclear experiment than had been invested in it.

Bohr sailed for America, still a neutral nation, and, a week after arrival, at a physics conference in Washington, reported the Meitner-Frisch formula. He cabled Meitner from America:

> "Just received copies of your's and Frisch's "Nature" notes. Heartiest congratulations on most important discovery. Best wishes for continuation of work on wonderful new phenomena."

The news swept through the physics community in the U.S. within days. An entire generation of young American chemists and physicists was thus prepped for the "Manhattan Project." When Japan drew America into the war, the best of them were recruited into the elite group that, within an amazingly short time, developed the A-bomb.

In England, physicist Sir Rudolf Peierls had almost immediately applied the fission knowledge to the secret nuclear-research MAUD atomic energy program in which he was employed. "There is no question," he wrote, "about the importance of the note by Meitner and Frisch, with its clear discussion of the physics of the fission process in simple terms."

A year later, World War II was on. Frisch was working in England with Peierls at the University of Birmingham. Based upon the Hahn-Strassmann experiments and the Meitner-Frisch hypothesis, they calculated that barely two pounds of uranium 235 would be enough to make an atomic bomb; they explained how the isotope uranium 235 could be extracted from the abundant 238; and, finally, how it could be detonated. This was the genesis of the Manhattan Project.

Hahn revealed the truth of Meitner's involvement in splitting the atom – backhandedly. He wrote to a colleague and expansively boasted "Only after several physicists had expressed their astonishment that slow neutrons should initiate two successive a-processes in uranium did Strassmann and I, in order to dispel the doubts of the physicists, investigate still more carefully the properties of our radium-isotopes." The "physicists" of course were Meitner and Frisch. Which he decided not to acknowledge, so he thereafter eliminated physics from further discussion. Through chemistry alone, he claimed he had invented the process. The Meitner/Frisch explanation of the physics involved in fission was shrugged off as little more than a postscript to his chemistry. Frisch speculated that Hahn was "annoyed that the proof for (fission) is so much simpler and easier when physicists analyze the data …"

The next move in fabricating a rewrite of history was not long in coming. Meitner's former student, Friedrich von Weizsacker, in a paper announcing his discovery of plutonium –an alternative fuel for an atomic bomb – reviewed the essential steps that led up to the nuclear fission discovery. Meitner was not cited by name. Her preliminary papers had to be listed (Hahn had never written any) – but again without mentioning her name.

In the fall of 1939 – almost immediately after the war commenced – a nuclear bomb program was initiated in Germany, whereas the atomic-bomb "Manhattan Project" in the U.S. did not commence until 1942. Nobel Laureate Enrico Fermi and his colleagues built an atomic pile in Soldier's Field, Chicago. December 1942, he produced the first self-sustaining chain reaction. Meitner, in Sweden, was urgently invited to join the team assembling in Los Alamos to build a nuclear explosive. But she had seen the carnage of war first hand, in 1914-18. "I will have nothing to do with a bomb!" she said.

English and American forces invaded France in June 1944, with air, sea, and land forces – but not yet nuclear explosives. German guided-missile V-bombs were reducing much of London to wasteland. Lise wrote to friends her fear that German rockets would soon be fitted with nuclear warheads. The same fear haunted General Leslie Groves, head of the Manhattan Project, and J. Robert Oppenheimer, its director. Hence, when allied forces were battling toward Berlin in 1945, a U.S. commando team raced south to where the German nuclear group had been relocated when allied bombing rubbled Berlin. It was still feared by the Allies that – at the last minute – an atomic warhead on the V-2 missiles being fired from eastern Germany would turn defeat into a stalemate, and would save the regime in Germany.

The commandos reached the nuclear research headquarters and learned there was no German A-bomb nearing completion. There was only a small-scale reactor for generation of power. To keep Germany's top scientists out of Russian hands – the Red Army was approaching – eight of them, including Hahn, were flown to England, and interned at Farm Hall, a secluded country estate near Cambridge. Their rooms and lounges were bugged; every word spoken was recorded.

When news arrived that an American uranium bomb with explosive power equal to 20,000 tons of TNT had leveled Hiroshima, they were stunned. The Germans simply could not believe that their scientists had been outclassed by those of another country. The British officer-in-charge of monitoring their conversations was astonished by what he termed their "inborn conceit." They were actually pleased that exiles from Germany had "set it all in motion!"

Von Weizsacker introduced an idea that became a mantra among his countrymen. "History will record that the Americans and English made a bomb

... (while) the Germans under the Hitler regime, produced (a workable reactor for) the peaceful development of ... (power from) the uranium machine ..." Otto Hahn pursued his personal agenda: "As long as Prof. Meitner was in Germany, the fission of uranium was ... considered impossible." (No argument there – it had been considered impossible by Fermi and every other investigator; and was "discovered" only after she left Germany.) He continued: "Based on extensive chemical investigations of the chemical elements which resulted from irradiating uranium with neutrons, Hahn and Strassmann ... (learned that) uranium splits into two pieces ..." This was a restatement of the position he took in 1939, asserting that he and Strassmann "never touched upon physics, but only did chemical separations over and over again." (No mention was made by Hahn, then or ever, that his chemical separations were so puzzling that he pleaded with Meitner not once, but twice, to explain why his experiments were failing.)

Upon this series of ingenuous misstatements, Hahn built his position, and stuck to it through his lifetime: that Meitner had nothing to do with the discovery of nuclear fission – solo credit was his. His Farm Hall colleagues soon echoed that "party line."

By the time the Farm Hall Eight were sent home, another catechism emerged, likening America's strike against an enemy country – in wartime – as parallel to their own country's sin in overrunning peacetime neighbors. Further, the Allied "occupation troops," they complained, were just as brutal as Wehrmacht and SS forces had been in conquered lands. Destruction of German cities, infrastructure and communications were vicious, beyond justification, they charged. For years, loud were the complaints about German lack of housing, fuel, food, coffee, cocoa, tobacco. Not till a new generation of young Germans, who inherited the devastation, were the first voices raised by writers and filmmakers, not scientists, identifying their parents as Hitler's willing accomplices. The scientists never did accept responsibility. To Otto Hahn, Lise Meitner wrote:

> "You all worked for Nazi Germany and you did not even try passive resistance ... millions of innocent people were murdered and there was no protest ... one path for you would be to deliver an open statement (now) that you are aware that, through your passivity, you share responsibility (when) you first betrayed your friends, then your men and your children ... in a criminal war ... and finally you betrayed Germany itself ... you had no sleepless nights, you did not want to see, it was too uncomfortable."

These blunt accusations justified, in Hahn's heart and mind, his efforts to deprive Meitner of any share of credit for their historic discovery.

155

The atomic bomb changed Lise Meitner's quiet life. Word spread that she was involved in building the weapon. When she refused interviews, Swedish journalists had no choice but to fabricate stories to go with sensational headlines. One for instance, lauded the "FLEEING JEWESS" for escaping Hitler with the secret of the bomb in her purse. Other writers used as research material a fanciful "Saturday Evening Post" article which described how Meitner invented the formula for fission on a train fleeing Germany; she became the "Jewish mother of the bomb."

But the 1944 Nobel Prize in chemistry was awarded to Otto Hahn. Meitner – if she were mentioned at all in the press releases – was described as an assistant, or as a subordinate. Many scientists were scandalized. Members of the Nobel advisory committee for physics protested vehemently. Hahn's work in chemistry meant little, unlinked to the physics supplied by Meitner. The Swedish Royal Academy for Science demanded a review. The Nobel award-grantors pointed out that, if Meitner were named, it would also be necessary to add the names of Fritz Strassmann and, of course, Meitner's nephew, Frisch. But – it was explained – Nobel rules prohibited more than three names per award. The decision to honor Hahn alone stood. Niels Bohr, himself a Nobel Laureate and thought by many a peer to Einstein, proposed that Meitner be named Nobel awardee in Physics the next year; he again proposed her in 1947 and 1948. The imbroglio had become ugly. Questions were asked why, with the war still on, Sweden's Nobel Foundation even considered honoring Hahn. Was it evidence that supposedly "neutral" Sweden had actually been a de-facto partner of Nazi Germany? The upshot was that an error had been made, and it would not be corrected.

Recent release of the till-now unpublished deliberations of the Nobel Committee make it clear that the jury did not understand that Meitner had worked on the nuclear experiments with Hahn and Strassmann before her flight to Holland, Denmark, and finally Sweden; nor that Meitner had guided Hahn and Strassmann through the meeting in Copenhagen and subsequent correspondence. (The archives further reveal that Meitner had previously been nominated for discoveries in the 20s and 30s.)

Snubbed in Sweden, Meitner was idolized in America. She was reunited with sisters in New York; her nephew came by train all the way from Los Alamos. The press hailed her as a "pioneer contributor to the atomic bomb." She was feted by professional societies and awarded honorary doctorates by universities. The National Press Club named her "Woman of the Year." President Truman attended the award banquet and sat beside her. She became a celebrity, was accosted on the streets, deluged by mail, courted by publishers, received a film studio offer to participate in "The Beginning of the End," a movie that would dramatize her flight from Germany with the bomb secret in her purse. She refused. The fee offer was raised. She still refused.

Meitner was back in Stockholm for the award of the Nobel Prize to Hahn. She met him at the train; they were photographed; she was identified in one newspaper as Hahn's "former pupil." Another newspaper gave her more credit: "... on hand was Prof. Lise Meitner, Prof. Hahn's world-famous pupil, whom he heartily embraced when she came to meet him." Interviewed often, he never acknowledged they had worked as partners, nor the part she played in the nuclear-fission discovery. He never mentioned her name in public appearances or press interviews. He used his "thank you" remarks at the Nobel ceremony to assert that Germany was

> "probably the most unfortunate country in the world ... (it is) really not true that all Germans, and especially the German scientists, subscribed to the Hitler regime ... (they had) no chance to form their own opinions, no free press, no foreign radio broadcasts ... people outside Germany (don't) know the oppression which most of us experienced for the last ten or twelve years."

In the autobiography written when he was in his 80s, Hahn notes with surprise that those close to Meitner had been "rather unfriendly ... after all, the prize was awarded to me just for work I did alone ... Lise Meitner received many honorary doctorates in the USA and ... was 'Woman of the Year.'" In his next public statement, Hahn amplified that nuclear fission was an achievement of radio-chemistry – made in spite of opposition by the nuclear physicists. Then he flogged the anti-American theme: he was glad "Germans were not burdened with the responsibility for the bomb and the resulting meaningless deaths of thousands of people ..."

The circular thinking that assuaged German guilt for the 13 years of the Hitler regime's crimes had embraced his own self-deception, like a worm in the heart of rotten fruit. It is pointless to wonder whether Hahn had self-doubts. Never in public, or in his autobiography, or in correspondence, or in conversations, did he flinch from his fiction. All but a few in German science joined in the sycophantic chorus that was born in Farm Hall. But those few dissenters included peers who knew the pair well during their decades together. Enough memoirs were written, enough reviews were published in scientific journals, to initiate rehabilitation of Lise Meitner's' reputation. After nine years on a pittance stipend, she was asked by the Swedish government to head a new nuclear science department at the salary of a full professor. She was greatly pleased, chuckling "Here, just being a woman is half a crime and ... having one's opinion is completely forbidden."

She was next offered an official appointment in Germany. Fritz Strassmann - who during his de-nazification hearing had firmly testified that credit for his and Hahn's fission success was due to Meitner's guidance - was by 1949 director of the physics institute transferred to the University of Mainz in divided Germany. He asked Meitner to become head of the chemistry section. She turned to her old friend, Hahn, for advice. Instead of directly answering her inquiry about the Mainz offer, he urged her not to

> "constantly reproach an entire people for their behavior ... We all know that Hitler was responsible for the war and the unspeakable misery all over the world, but there must be some sort of world understanding also for the German people."

She also asked advice from Max von Laue, another friend and Nobel Laureate, whom she had considered to be fair-minded and trustworthy, all through the dark years. She wrote her concern about working with German assistants, who would not be friendly "in part because I am Austrian, and in part because of my Jewish origins." The trustworthy friend reassured her that Germans had "no objections to Austrians."

She declined a publisher's request for her autobiography, saying such books were "either insincere or tactless, usually both." Awards, prizes, honorary doctorates, medals showered upon her. In 1964, the U.S. Atomic Energy Commission offered the distinguished Enrico Fermi Prize to Meitner. Hahn, when also asked, nominated Strassmann. Ultimately, the award went to all three, Meitner, Hahn and Strassmann. It need be noted that, though honor came late to Strassmann, he had earned it early by refusing to work for the nazis in the 1930s – though he was Aryan and unemployed; and he and his wife had sheltered a Jewish pianist, Andrea Wolffenstein, in their apartment during the war.

Hahn meanwhile had become world-famous. He was pictured on postage stamps; his name was put on buildings, institutes, libraries, streets, schools, roads, trains, bridges, coins, an island in the Antarctic. He was honored by scientific societies all over the world. In Berlin, when the partners' old laboratories were rebuilt, the structure was named for Otto Hahn, with a bronze plaque for him and Fritz Strassmann. No mention was made then, or since, of Lise Meitner.

But evidence mounted steadily that a great injustice had been done. When a new facility for nuclear research was planned in suburban Wannsee, it was proposed to name it after Meitner. By then, it was inconceivable to honor only one, not both the partners. So the sign above the security gates of a handsome campus – originally intended to honor only Meitner – now reads "Hahn-Meitner Institute."

Meitner moved to Cambridge in 1960, to spend her last years near her nephew, Otto Frisch, who was a faculty member there. She died just short of 90.

FLORENCE NIGHTINGALE
Triumphed over a
Gaggle of Goliaths

Part One: England

There is still room for improvement ... as regards the modesty and obedience, of nurses; ... Nurses must always be servants, and they cannot ... be permitted to rise above that position in society. The ... more respectable they are the better, provided always that their respectability does not interfere with their obedience.

Thus spake "The Lancet," Britain's medical journal, in its issue of June 4, 1881. Crude male chauvinism? So thought Ethel Gordon Fenwick, founder of the British Nurses Association when she spat "the nurse question is the woman question, pure and simple." But there was far more to it than that.

The extraordinary career of Florence Nightingale exemplifies. She was born in 1829 into country gentry. Her family, like all families of their class, wrapped their young women in a silken web, trained them for household management; "proper" manners, social graces, riding and shooting, criteria for choosing a mate, motherhood and the code by which their children were to be reared. They were taught foreign languages, needlework, flower arranging, dancing, musical instruments, gardening, "putting up" preserves.

Florence and her younger sister were served by maids from the moment they awoke, to the moment they said their prayers at night. Her father never worked a day in his life; his major achievement was inheriting wealth from his uncle in exchange for abandoning his father's surname and taking on that of the relative.

The girls grew up surrounded by aunts, uncles, first, second and further removed cousins — more than 50 of them by the time they were in early teens. The calendar was crowded with happy events in that closed world: marriages, anniversaries, births, First Communions, entertainments, shopping for new furniture, Turkish rugs, Venetian chandeliers, enlargement of their homes, enhancement of their gardens. Florence had a pleasure - loving nature and an enormous vitality, could dance all night. Lord Byron's daughter, Lady Lovelace, wrote poetry about Florence's "soft and silver voice" and her "grave and lucid eye." She was home -taught to read and speak most European languages, and at ease in the highest levels of British society; more than once took tea with Queen Victoria, was friendly with the prime minister, royal highnesses, dukes and lesser nobility – even the future Napoleon III. She was popular with people of all ages.

Outwardly, she was a typical representative of her class.. Inwardly, she suffered deep discontent, particularly as she advanced into her 20's. The very qualities that earned popularity in social life — energy and intelligence — eventually made her dissatisfied with it. "I craved for some regular occupation, for something worth doing instead of frittering time away on useless trifles."

But boredom does not, of itself, induce action. That evolved gradually, as she accompanied her mother in "squire's wife" duties, clattering in a carriage around the extensive Nightingale estate, visiting sick peasants in their cottages, distributing money and soup. She became accustomed to congestion, slatternly ignorance of sanitation, dirt, drunkenness, domestic brutality. She noted in her diary "My mind is absorbed with the idea of the sufferings of man ... I can hardly see anything else and all that poets sing of the glories of this world seems to me untrue. All the people I see are eaten up with care or poverty or disease." As she grew older, she slipped away to visit the cottages alone, on foot. She became devoted to caring for others. Finally, she expressed her leanings to the Chevalier Bunsen, Prussian ambassador, who was a houseguest. "What can an individual do towards lifting the load of suffering from the helpless and miserable," she asked. He recommended an institution at Kaiserswerth in Germany where "young women of quality" were trained to become deaconesses in service to orphans, the poor, the aged, the sick.

But at 22, Florence was not yet ready. Two years passed until another set of guests, socially-enlightened Americans, visited the Nightingales. Dr. Gridley Howe and his wife, Julia Ward Howe, (known now as the originator of "Mother's Day), were dedicated philanthropists. They had founded an institution for the blind — first of its kind in the world. They were planning a program to provide free medical care to the aged and the poor. Florence asked, "Dr. Howe, do you think it would be unsuitable and unbecoming for a young Englishwoman to devote herself to works of charity in hospitals and elsewhere? ... Do you think it would be a dreadful thing?" Howe's answer: " ... it would be unusual , and in England whatever is unusual is thought to be unsuitable; but I say to you ... if you have a vocation for that way of life, act up to your inspiration and you will find there is never anything unbecoming or unladylike in doing your duty for the good of others ... God be with you." Dr. Howe was so deeply impressed by the idealism of the young woman that when a daughter was born to his family, she was christened Florence.

The phrase "in England whatever is unusual ... is thought to be unsuitable" needs explanation. It was unsuitable for a female to be interested in mathematics, for instance. Florence was brilliant in that subject and wanted to study it seriously. All tutors were male, however, so chaperones had to be interviewed. It was a burden for her parents to deal with such an awkward problem; the ambition was shelved. Florence had, however, mastered the mysteries of math sufficiently to tutor a young male friend of the family for his entrance exams at Sandhurst, the West Point of England. When this innocent misdemeanor became known, it caused a scandal in their social circle — the boy would be laughed out of the army when it was learned that his tutor in math was a woman!

Despite these experiences, and despite the respect she bore for her family, came the day when she declared her wish to become a nurse. Her mother wailed that the girl had "an attachment of which she was ashamed ... some low, vulgar surgeon" and would "disgrace herself." Her father was shocked at the waste of her talents, the waste of teaching her Greek and Latin, French and Italian, poetry and philosophy. The violence of their emotions need be understood in context of the times: hospitals, she later wrote,

> "owing to lack of cleaning and lack of sanitary conveniences for the patients' use, had become saturated with organic matter ... Walls and ceilings were ... also saturated with impurity ... windows were kept closed for warmth ... (or) boarded up in winter. After a time the smell became sickening, walls streamed with moisture, and ... vegetation appeared ... The beds on which the patients lay were dirty. It was common practice to put a new patient into the same sheets used by the last occupant of the bed, and mattresses were generally ... sodden and seldom if ever cleaned ... The nurses ... slept in wooden cages on the landing places outside the doors of the wards, ... where there was not light or air ..."

And with whom would Florence consort?

"It was preferred," she wrote, " that the nurses should be women who had lost their character ... have had one child." Said a physician "The nurses are all drunkards ..." Echoed the head nurse at another hospital, "in the course of (my) large experience (I have) never known a nurse who was not drunken ... immoral conduct (is) practiced in the wards."

The idea that a young woman should *WANT* to become part of that world was unthinkable. In Florence's family circle, there was shock, anger, shame. Mother and sister resorted, almost daily, to hysterics. In the face of this household tumult, Florence dithered. But her normal social life continued to be distasteful. What alternatives were there? Only marriage. Florence had her opportunities — and she enjoyed the company of men. So why didn't she wed? This is her assessment of one wealthy and popular suitor, highly popular with her parents and friends:

"I have an intellectual nature which requires satisfaction and that I would find it in him. I have a passionate nature which requires satisfaction and that I would find it in him. I have a moral, an active nature which requires satisfaction and that I would not find in his life ... making society and arranging domestic things."

She refused him, and suffered. "I know that if I were to see him again ... the very thought of doing so quite overwhelms me ... not one day has passed without my thinking of him ... life is desolate ..."

While she waited for "Mr Right" to enter her life, Florence became obsessed with the conviction that her real, true, destiny was to become a nurse. Stony disapproval not only came from her father, mother, sister, but also from friends, the family doctor, both Church of England and Roman Catholic clergy. Religion, itself, did not attract because it offered men "bishoprics, archbishoprics, and a little work ..." whereas for females it recommended "crochet(ing) in my mother's drawing room; or, if I were tired of that, to marry and look well at the head of my husband's table." Her aunt urged this goal: "a house-party or a dinner-party can be understood as a "glory of God." (Entertaining dozens of weekend guests in holiday season was routine for the Nightingale women.)

The Roman Catholic Church offered more practical opportunities for young women, and at one time Florence considered conversion. She explained her goals and outlined her faith in great detail to Archdeacon Manning (later cardinal). But as Cecil Woodham Smith relates in his classic biography of Florence Nightingale, "she did not touch on the Christian doctrines of salvation, redemption and the incarnation of Christ ... her God was God the Father, not God the Son ... he refused to accept her as a convert."

At 32, Florence went to Paris, ostensibly to visit with one of her oldest friends. Through "connections," she procured a permit that provided entry to every hospital, also infirmaries, almshouses, orphanages, other institutions. She watched doctors perform operations; assembled "tables of organization," circulated questionnaires all over Europe, collected statistics, read mountains of dry-as-dust reports. She applied — and was accepted — to become a trainee in the hospital of the Sisters of Charity. Whereupon her aged grandmother took to bed with what proved to be her final illness — and Florence was called home to help with the nursing. When the end came, family demands swept her up into the old routine. Her entire life might have continued in that pattern, as it did for the millions of single young women, except for the intervention of fate.

Years before, she had met, in Rome, Elizabeth Herbert, newly wed to Sidney Herbert, master of Wilton, one of England's glorious stately homes. He had great wealth, great charm, great ability as a Member of Parliament. Florence grew closer with them in the years that followed. One special day, Elizabeth wrote to her young friend that, with other titled Ladies, she was "on the Board" of an "Institution for the care of Sick Gentlewomen in Distressed Circumstances" in London. It was in desperate financial condition, and needed a Superintendent to undertake reorganization. Florence applied for the post, even though it was unpaid. Not only unpaid, but she was expected to bring to the institution an assistant of mature years, to compensate for her own youth — she was still in her thirties — and to pay the assistant out of her own pocket. She accepted these terms in exchange for something more important to her — supreme authority, freedom even from answering to the "board" of Ladies and their titled gentlemen. It took months of negotiation before these terms were agreed to, a battle royal.

"There is ... much jealousy in the (sponsoring) Committees," she wrote to her father. In consequence, she had to learn the fundamentals of business tactics. "I do all my business by intrigue," she wrote. She played off entire Committees against each other. "Last General Committee I executed a series of resolutions ... and presented them as coming from the Medical Men." When the lay Committees approved, "I showed them to the Medical Men ... (who) approved them all ... (thinking) they were their own!" Only once did she enter into direct battle, on a matter of principle. "My Committee refused me to take in Catholic patients," she wrote to a friend, whereupon she threatened to quit "unless I might take in Jews and their Rabbis to attend them. So now ... we are to take in all denominations ... to be visited by their respective (clergy)." Of course, it was expected that I "take him upstairs myself, remain while he is conferring with his patient, make myself responsible that he does not speak to, or look at, anyone else, and bring him downstairs again ... and out into the street ... "

Much else had to be learned. Those who were supposed to be responsible for financial management, hiring of personnel, procurement of supplies were at best ineffectual, at worst, incompetent. She gradually took over all these functions. She haggled wholesale contracts for provisions and drugs. She left no detail to chance. "The chemists sent me a bottle of ether labeled Spirits of Nitre, which, if I had not smelt it, I should certainly have administered ... (it, resulting in) poisoning," she wrote to her father.

Within six months, Florence Nightingale had whipped her Institution into shape, and her reputation for efficient administration spread through London. Sidney Herbert was now a government minister. He asked for data on hospitals, on nurses, on their pay and housing. From statistics gathered over the years, Florence was able to fulfill every request. With her Institution ticking along like a well-oiled Swiss timepiece, she had no hesitation, when cholera broke out in the "red light" district of London, volunteering to superintend the nursing at the local hospital. So many nurses had died of the disease, and so many fled in fear of infection, that she was almost alone in tending to the prostitutes. This came to the notice of a doctor, who asked if she would consider "taking over" as

Superintendent of Nurses at Kings College Hospital – and promised that England's first training school for nurses would be created for her there, to establish the vocation on a professional basis.

The Kings College appointment didn't come through in time. England and France, allied with Turkey, declared war on Russia. Forty years after defeating Napoleon at Waterloo, Britain was still intoxicated with belief that its army was invincible. Gorgeously -attired Guards regiments, before embarking for the threat of war, paraded through London, headed by drummer boys and blaring bands. Sidney Herbert had been appointed Secretary at War. He wrote to Florence Nightingale:

> "none but male nurses having ever been admitted to military hospitals ... no military reason exists against ... introduction of women ... There is but one person in England who would be capable of organizing and superintending such a scheme ... you."

His letter to Florence crossed in the mail with her letter to his wife, Elizabeth: "A small private expedition of nurses has been organized ... I do believe we may be of use to the poor wounded wretches." Would her husband expect the scheme to be countenanced by the authorities? "Would he give us any advice or letters of recommendation? And are there any stores for the Hospital he would advise us to take out?" Finally, she asked her friend to write to the wife of the British ambassador in Turkey and assure her "This is not a Lady but a real Hospital Nurse ... (with) experience."

Events moved swiftly. Sidney Herbert nominated her to become "Superintendent of the Female Nursing Establishment of the English General Hospitals in Turkey." Support for her appointment came from the highest levels, and Parliament endorsed it unanimously. All England applauded, when the news was reported. Never before had a woman been officially so honored.

Part Two: The War

The battle-zone was in Russian Crimea. Field hospitals were established around Sebastopol, but the major general hospital would be organized at Turkish Army barracks in Scutari, a village on the Asian side of the Bosphorus Straits, reachable from the war zone by boat. Too few vessels were available, so transport officials allowed only soldiers on board. No food or cooking supplies, no tents, no bandages, no splints, no chloroform, no drugs or medicines of any kind, no surgical instruments or other hospital equipment, no litters, no carts to carry the wounded, no bedding, no food, no sanitary supplies. Untreated, sick and wounded men were packed like sardines into small, leaky tubs for transport to Scutari. Screaming in pain, they were unable to crawl to sanitary facilities, so they rolled in mounds of filth as the ships tossed in the waves. Unloaded at Scutari, they were dumped on the ground; bedding was straw stiffened by manure. No operating tables; no tables or chairs of any kind.

None of this reached the British public, which exulted in news of glorious victories — until a man named William Harris Russell was sent by the London Times to the army base and trenches at Sebastopol. The first-ever war correspondent, he exposed an awful truth: "no sufficient preparations have been made for the care of the wounded ... there are not sufficient surgeons ... dressers and nurses ... there is not even linen to make bandages ... the men (are) kept, in some cases for a week ... left to expire in agony, unheeded ..."

In London, Florence had been told by the chief medical officer of the army there were more than ample supplies at the battlefield and at Scutari. Massive shipments were enroute already to create surplus he said. She was told more stores were not needed. But she knew about the male ego, however, so when her ship docked for four days at Marseilles, she bought supplies that she knew would be needed in every sickroom.

Meanwhile, cholera struck at Scutari. The medical staff was overwhelmed. Sick and wounded soldiers lay everywhere, half-naked. In a panicked battlefield retreat, they'd been ordered to abandon kit-bags. Army regulations prohibited issuing fresh clothing to men who were careless with their original issue. Army regulations assumed that the first issue of eating utensils would also last a lifetime; under no circumstances could "replacements" be issued. Army regulations overwhelmed the medical staff with paperwork. Army regulations, as always, in every war, represent an implacable enemy to the ordinary soldier.

With the hospital bursting at its seams, the adjoining barracks themselves – four miles of corridors! – were ordered to be converted into hospital wards. Army orders failed to identify who was responsible to implement those orders. During an official cabinet-level Enquiry of war-zone bungling, it was repeatedly argued by senior officers that their duty was to issue orders, not to oversee how – or if – they were being carried out. (Lest it be thought that this witlessness was peculiarly British, consider the number of American soldiers who died of exposure – froze to death at their defense posts in late December – during the Battle of the Bulge in World War II because they had not yet been issued winter clothing. The same idiocy was evident in the Wehrmacht, during its headlong retreat from Leningrad and Moscow. Napoleon's commanders had been just as negligent.)

Even more evil was the contempt upper class officers demonstrated for lower-class men under their command. The Duke of Wellington, victor over Napoleon at the Battle of Waterloo a generation before, memorialized this attitude by describing the rank and file of his army – whose courage had routed the most successful military machine of modern times – as "the scum of the earth enlisted for drink." A regimental doctor during the Crimean campaign testified that the traditional class-consciousness of English society was unchanged: " ... no general officer has visited my hospital nor ... interested himself about the sick." Another physician, veteran of the action before Sebastopol, used the words "apathy and indifference" in describing officer attitudes toward their foot-soldiers. He bitterly noted that "very tolerable (housing) ... has been raised for the occupation of (high-ranking) individuals while men labouring under disease are left on the damp ground in a leaky tent."

Stories like these are included in this account only because they have direct bearing upon the situation into which Florence Nightingale and her 40 volunteers were plunged when they arrived November 4, 1854. They were assigned four rooms, a large closet, and a kitchen — filthy, damp, unfurnished, innocent of beds, innocent of bedding. There was no food, no cooking implements. There was, however, a strong smell. Tracing its source, they went upstairs and found the corpse of a Russian general, forgotten in his cell. They dragged it out, but could find no soap or disinfectant — not even a broom — to scrub down the room. Welcome, girls.

Next morning, having slept in their clothes, on the floor, they reported for work – and were rejected. Florence was sternly warned not to "spoil the brutes ..." further described as "animals" and "blackguards" and "scum." Everything possible was done to dissuade the Nightingale nurses from interfering with routine medical practice in the Scutari horror. Doctors and army officers alike assumed Florence was a spy for the civilians "back home." In a sense, this was true — Sidney Herbert had specifically enjoined her to write privately to him. But that was not her main mission. Relief of suffering was why she'd come. She had to devise tactics appropriate to the adversaries – as she had at her Institution in London.

Her choice was drawn from Greek folklore: feminine passive resistance. Nightingale women were not allowed to enter a ward without invitation from a doctor – and no doctor issued such an invitation. So Florence confined them to their rooms, kept them out of the sight of men. Hardest of all was persuading nurses to ignore the screams of patients in atrocious pain. She kept them busy sorting and washing old textiles, stripping them into lengths suitable for bandages. They scrubbed, disinfected walls and floors using supplies bought in Marseilles.

Until yet another overloaded ship landed at Scutari, piled high with desperately sick, gravely wounded men from the disastrous rout of British forces at Balaclava. There was no food available for them in the hospital kitchen. There was food available in Miss Nightingale's kitchen. Moved to mercy, the doctors authorized meal preparation. The first breach in the stone wall. It widened, and then the wall itself crumbled when, November 9, evidence arrived that disintegration of the British Army of the Crimea was in progress. Pinned to defense positions that were indefensible, short of men, guns, ammunition, food, clothing, medical supplies, every battle with the Russians ended in disaster. Hundreds of brave soldiers were felled by the improvidence and arrogance of their commanders. Hellishly suffering men awaiting transport had amputations at the harbor — arms, legs, entire torsos clogged the water. When those still alive were finally loaded on the ships and dumped ashore at Scutari, the already demoralized staff there had to deal with a tidal wave of cholera patients, erisipelas, fever, gangrene, starvation, hypothermia, scurvy, frostbite. Civilians were pressed into service as surgical assistants, becoming human operating tables by reaching under and gripping the wrists of a partner on the other side of the thrashing patient. Their four arms, hands clasped, made possible the surgery – without anesthetics or grog. Those awaiting their turn were crawling with lice and flies, lying in filth. Ordure floated on all floors an inch thick. The men relieved themselves where they lay or stood; food was dumped at their sides, splashing in the sludge. 1,000 men writhed in acute diarrhea; waiting for a chamber pot – only 20 were in circulation. 50% of the men who arrived in Scutari alive, died there.

Finally, Miss Nightingale and her nurses were sent for and started to do that for which they had come. But then ...

November 14, the worst hurricane on record hit the Crimea. It destroyed everything in its path — buildings, tents, vessels in the harbor being loaded or unloaded – they all sank in minutes. Army administration fell apart. A visiting member of Parliament, Augustus Stafford, present on an "inspection tour," testified that from top to bottom "there was a kind of paralysis, a fear of incurring any responsibility, and a fear of going beyond their responsibilities." When he asked the commandant of the hospital why the lavatories were useless, the senior official answered that his instructions did not cover such work. The civilian offered to pay laborers' wages out of his own pocket. He was refused; the army would accept no help from civilians.

But Miss Nightingale had money and authority directly from the Parliament. So, from that point on, whenever and whatever the military quartermaster could not — or would not — provide, the request went to Florence. She sent agents into Constantinople, then the world's largest shopping center, to buy knives, forks, socks, underwear, shirts, tin cups, spoons, trays, tables, scrub-brushes, brooms, bandage linens, tables and chairs, lye, towels and soap, food, combs, disinfectant, vermin-poison, tin baths, combs, scissors, bed-pans, pillows for stump-rests. It was all available. It had always been available.

After the storm, a motley fleet of fishing boats, cargo tugs, ferries and whatnot was pressed into service as hospital ships. They had no facilities to tend sick or wounded men. At Scutari, the flotsam was dumped on the dock — hundreds more wounded, hundreds more sick, hundreds more amputees, naked, filthy, starved, emaciated. Disused barracks had to become hospital wards overnight. The Turkish workmen went on strike for higher wages. Miss Nightingale fired them all, hired a new bunch — twice as many, paid for with private funds. She had pallets ready for 1,000 more patients before they arrived. The "Times" correspondent quoted one of the exhausted Balaclava veterans who "thought ... we were in heaven!"

That was the high point of Florence Nightingale's career in the Crimea.

It is one of man's perversities to surrender to envy, to jealousy, to power struggles, to pander to pecking order, to kowtow to power echelons, to class, professional, and gender discrimination. Competent nurses raged that they were prevented by doctors' orders from determining themselves how much food a desperately sick patient needed. Incompetent ones whimpered against Miss Nightingale's iron discipline, her control over their religious convictions when they interfered with care of patients. She brooked no arguments: sent back to England those who got drunk regularly and were promiscuous. Nurses who were members of Catholic Orders insisted on recognizing the authority only of their Superiors in the church. London newspapers were peppered with letters that accused Miss Nightingale of complicity in a "Romanist establishment;" or, alternatively, being a tool of "Anglican Papists" in the War Office; or even of involvement in "Jesuit conspiracies." To separate the feuding sectarians, some Irish nuns were sent home — which of course enraged the R.C. population in England. Not to be left aloof from the fray, many Protestant parsons warned their constituencies against sending aid supplies to hospital at Scutari.

The human heart is vulnerable to unworthy emotions. Old friends, thirsting for recognition, became Florence's rivals. Some friends who remained faithful, wearied of the internecine strife, and announced their leave-taking. Meanwhile, the deluge of sick and wounded continued without letup. 4,000 more broken men arrived in the weeks following the hurricane; the death rate climbed. Miss Nightingale's chief aide, Charles H. Bracebridge, wrote to Sidney Herbert that "Flo has been working herself to death, never sits down ...without interruption ... the attempt to do more will kill her ... today 200 sick landed looking worse than any others yet."

It didn't kill, but weaken her it did. After long hours of "rounds" in the wards; after abrasive sessions adjudicating complaints of nurses, doctors, chaplains, merchants and army functionaries; after supervising the preparation of hundreds of patients' meals; she worked far into the night – writing last messages to the families of boys who died, holding her hand; and warm notes to the families of nurses, who often had inadequate education to write themselves. She acknowledged each and every package that arrived from England, however useless, confiding to her diary: "There is not a small town, not a parish ... from which we have not received contributions ... not one of these is worth its freight, but the smaller the value, of course, the greater the importance the contributors attach to it."

Sanitary conditions were impossible to correct, even to improve much. Underneath the hospital was a vast cesspool of human waste, rotting amputated limbs. Inside the hospital, the porous plaster walls became drenched in poisonous gasses. Rats scurried everywhere, grew fat. Doctors refused even to enter wards where cholera victims were dying; they shouted their instructions to nurses, inside, through closed doors. Florence wrote to London: "Matters are worse than they were two months ago, and will be worse two months hence than they are now." She developed a monumental contempt for army officialdom. Sidney Herbert thus learned that this or that a senior officer was "an insincere animal at the bottom ... neither gentlemen nor men of education, nor men of business, nor men of feeling, whose only object is to keep themselves out of blame." Whereas her affection was boundless for the common soldiers who "amid scenes of loathsome disease and death, (evidenced) innate dignity, gentleness, and chivalry ... shining in the midst of what must be considered the lowest sinks of human misery..." Her affection, her compassion, knew no bounds. When others were asleep, it was a common sight for the "lady with the lamp" silently to walk the wards, stopping from time to time to touch and murmur comfort to a moaning man. Legend has it that men turned to the wall and kissed her slowly passing shadow. Her devotion was broadcast in England, by families of men who wrote from Scutari, to their neighbors and families. She became known as "The Nightingale in the East." Songs were written about her. Poems, too, including Longfellow's "Santa Filomena." Ships were named after her — and not only ships. The "racing news" section of a newspaper reported that "The Forest Plate Handicap was won by Miss Nightingale beating ... nine others."

She opened a recreation center for convalescing men; and writing rooms. She organized classes. She became the "bank" in which men deposited their pay, not to squander it on drink. She pressed into service the idle women of senior officers as teachers in classes to help soldiers acquire a degree of literacy. Nothing was too small a matter to deserve her attention – if it had to do with the welfare of the men.

The background of that picture was growing darker and darker. From the battlefield, William Howard Russell had been writing cablegrams to the London Times about continuing shortages of supplies. Though huge shipments had been sent, much was lost through thievery, but much more was lost through administrative incompetence. Shiploads of food were thrown into the harbor at Balaclava because they were not consigned to a specific person or office. Lime juice (an essential diet-supplement to prevent scurvy) was not issued for months – though a huge supply was on hand – because doctors had failed to prescribe it as part of the diet. More than four out of every five patients who reached Scutari were suffering from scurvy, painful and debilitating. Sick and wounded men lay in mud there because army regulations prohibited issuance of new blankets –

mountains of which were available – in replacement for those lost in battle. Russell gave an example: 27,000 shirts were thus withheld from distribution to shirtless men.. Fortunes were being made in England by manufacturers of supplies for the army: cots without legs; boots too small to be worn by boys. Survivors of (Kipling's) "The Charge of the Light Brigade" at Balaclava were dying of hunger and exposure. Their horses, too, starved to death. 11,000 "able bodied" men were all that remained before Sebastopol; 12,000 were bedded at Scutari; the rest of the Army of the Crimea was dead.

As is often the case, senior officers responsible for the debacle were promoted and transferred elsewhere – or honorably retired. Dr. John Hall arrived, newly-appointed Chief of the Medical Staff of the British Expeditionary Force. He immediately became a relentless, vicious, malicious enemy. Powerful as were Miss Nightingale's backers in London, they were no match for the network of influential friends he had acquired during 30 years of postings throughout the British Empire, men who now controlled army administration in London. (It had taken him 30 years to become qualified as a physician.) He was a man's man — spurned the use of chloroform! He instructed all officers under his new command to use, instead "the knife as a powerful stimulant ... it is much better to hear a man bawl lustily ..." Hall ruled his domain with terror, and went to any length to inflict revenge upon anybody who crossed him. Most of all, he hated meddling civilians, and never acknowledged that Miss Nightingale or her staff and her nurses and her purse had been of any value at Scutari. He refused to recognize her authority even to be there. His people, paralyzed by fear of Hall, translated the new climate into insolence toward Miss Nightingale and refusal to obey her orders. The simplest requests for supplies and procedures to alleviate the suffering of men were ignored. Hall demanded that all her requisitions had to be sent to him, personally, to be acted upon. He endorsed none. When a larger diet kitchen was at last put into service – Florence had designed and paid for it months before – 24 hours elapsed before officers' toast could be prepared in the new equipment. Hall complained officially, in writing, that Miss Nightingale was discriminating against his officers. She was required to file a defense: "for seven months ... every egg, every bit of butter, jelly, ale and Eau de Cologne which the sick officers have had has been provided (from) ... private pockets" (hers). Augustus Stafford wrote: "The nature of her difficulties is not understood and perhaps never will be."

Miss Nightingale collapsed, delirious, wavering between life and death for days, weeks. Dr. Hall tried to arrange transport for her and her party back to England. She discovered his intention when she was aboard the vessel – having been told a falsehood to carry her there. She had her stretcher taken off. Soldiers who could walk, drew lots for the honor of carrying her stretcher. Hundreds followed, silent, in tears. Convalescence took months.

Thus a full year in the Crimea dragged to its sorry end; and the Crimean War itself exhausted its energy to inflict damage on either adversary.

Dr. Hall continued his petty, malicious – and destructive – persecutions. He disputed Florence's nursing assignments. He transferred some of her most competent women from Scutari, where the need was great, to Balaclava, where officers were mainly cared for. He stalled in fulfilling her requisitions for supplies; on one occasion, he even withheld food. She finally wrote to Sidney Herbert: "We have now been ten days without rations ... during these ten days, I have fed and warmed ... at my own private expense by my own private exertions. I have never been off my horse until 9 or 10 at night."

The temptation exists to discount these complaints – pressures, overwork, sleeplessness, gender psychology and biology. If one is minded to find explanations, a dozen come to mind. What made them all frivolous and irrelevant were dispatches from the TIMES correspondent; eyewitness reports by Members of Parliament; and finally, undercover agents sent by the War Minister who investigated privately. All confirmed the vendetta, added details. The war-within-a-war exemplified the wrong-headed bungling incompetence, irresponsibility, and insensitivity of senior officers. All for the mean purpose of getting rid of the uppity female! The straws overloaded the camel's back; the War Minister became disgusted and published the following as a General Order, disseminated to all forces:

> "Medical authorities of the Army do not correctly comprehend Miss Nightingale's position ... Miss Nightingale is recognized by Her Majesty's Government as the General Superintendent of the Female Nursing Establishment of the military hospitals of the Army ... The Principal Medical Officer will communicate with the Female Nursing Establishment, and will give his directions through that lady."

This was far more than a slap on the wrist – it was a kick in the groin. It went on the personnel record of Dr. Hall. When, by seniority, he was slated for appointment in the top job in the medical corps of the British Army – by then "Sir John Hall" – he was passed over. No greater humiliation can be imagined in military life. Except for biographies of Florence Nightingale, he has disappeared even from English history.

Weakened in body, but greatly strengthened by the realization that no mountain was too high for her to surmount, Florence returned to England.

EPILOGUE

She declined the offer of transport from Constantinople on a British battleship. She declined an honor guard. She traveled incognito in a commercial vessel to France and from there, still incognito, to England. Careworn, emaciated, she was not recognized upon arrival, was not recognized when she entrained for home. She'd not arranged to be met; walked miles from the rail station to the family home and went into seclusion.

Rested, she undertook the campaign that revolutionized hospitals worldwide; and reformed army life for the ordinary soldier. As a first step, she went to Balmoral Castle in Scotland, Queen Victoria's home away from London. There followed a series of sessions. Warmly supported by the queen's husband, Prince Albert, the two women mapped a strategy, built upon Florence's personal knowledge and experience, her resolution and tenacity, the reverence in which she was held by the masses. Their objective was a Royal Commission, to be authorized by the Parliament, thoroughly to document the tragedy at Scutari and Balaclava; and set an agency for specific, item-by-item changes.

After every war, anywhere, matters relating to that war are pushed into the background. It wasn't allowed to happen this time. A huge ceremony was authorized by the Duke of Cambridge, Commander-in-Chief of the Army, in honor of Miss Nightingale. Great aristocrats vied for inclusion in the sponsoring committee; prominent politicians begged to be seated at the speakers' table. A Florence Nightingale fund was launched. Contributions poured in from all over the country. Thus was created a grassroots constituency for health-care reform.

Their national heroine became a magnet around which formed a group of right-minded people in the highest ranks of army, government, nobility. Pressure toward reform – the vehicle would be the Royal Commission – built steadily. Resistance was stubborn by Army and government bureaucrats. True then and true today: the principle that if no action is taken no consequences can ensue and no blame can be assigned to any individual. One of Florence's top allies, a former cabinet minister, defined the man who replaced him in office as handling a staggering workload without problem via "the simple process of never attempting to do it."

However senior, people like him were allowed no peace. They were bombarded by letters, visited by notables, pushed, prodded, cajoled, badgered and bullied. The power of the press was mobilized: journalists were cultivated by "leaks" to the newspapers and magazines that co-operated. Florence herself authored a pamphlet which proved to be one of the most damning. "Mortality in the British Army" was buttressed with facts and figures, diagrams, pie-charts – the first time this now-familiar "presentation" technique was used. She astounded readers by proving –– with statistics – that the death rate in peacetime, in ordinary barracks life, was far greater than in nearby residential communities. Every family in the country that ever had a son in service, or would in the future, became familier with the slogan "Our soldiers enlist to death – in the barracks!"

Nevertheless, the going was slow. In her diary notation at Christmas, 1856, Florence lamented that the "movers and shakers" in the British government and army didn't want to be moved or even shaken. In black discouragement, she concluded that her opponents were "generous with words about the Crimean disaster, "but the real tragedy began when it was over."

Six months dragged by. It began to seem that the Royal Commission would never be called for, never be authorized – even with support of the queen and pressure from leading political liberals. She set her jaw and issued an ultimatum. Sidney Herbert was out of office, but still one of the most powerful politicians in Britain. She threatened "three months from this day I publish my experiences of the Crimea Campaign —- unless there has been a fair and tangible pledge by that time for reform." He saw to it that Florence's warning got circulated in the corridors of government officialdom. That did it.

Within the deadline she had set, a Royal Warrant for the Reform Commission was issued. Hearings in Parliament began. Witnesses were called; made statements; answered questions. Miss Nightingale, of course, had hand-picked them, and provided each of them with the facts and statistics that brooked no defense. She testified herself in quiet voice; the Parliament sat silent, in shame. The victory was complete. The Report of the Royal Sanitary Commission on the Health of the Army was approved without dissent and ordered to be put into action.

That is not the end of the story, of course. More than half of Florence's 90 years were still to be lived; battles to be fought, defeats endured, victories celebrated. Reforms were made and unmade. Florence's ideas for how a modern hospital should be designed and functions were sometimes adopted, sometimes not. Some of Florence's demands were implemented by the army in the Indian Wars; some were not. She prevented a repeat of the Crimea Catastrophe by providing a scenario of the impossibility of supporting an English Army in the frozen, endless wastelands of North America; there was a powerful faction that wanted to headquarter an expeditionary force in Canada to widen the American Civil War by attacking the North as allies of the Confederate States. She won: the Cabinet retreated from that bizarre bellicosity.

She created the curriculum for professional training of civilian nurses, and founded the Florence Nightingale School at St. Thomas Hospital in London. Clones followed, in other cities. She demanded sanitary regulations be strengthened in the mushrooming cities. "Public health" became a prime concern on the national agenda.

Today, health-care professionals everywhere in the world know her story. Nobody has ever disputed the permanent significance of her work or her importance in history.   She accelerated social progress – without an army, without an organization, without "power" as the word is normally understood. Parallels are few in history.

HAROLD RIDLEY
Survived ridicule
To achieve fame

Harold Ridley "improved the quality of life for more people worldwide than any man alive" say those who award prizes for medical discoveries. More than thirty-five million people, in fact – and the total climbs daily. People with cataracts in an eye have had sight restored by the "Ridley Miracle" – an intra-ocular lens implant following removal of the cataract.

Yet Ridley's name is almost unknown. The taxi driver with whom I rode to Ridley's home from the railway station in Salisbury, England, didn't know the name. He didn't know that the lens implanted in his own eye – only months before – was the creation of a fellow townsman - for over 50 years. By nature a self-effacing man, Ridley's audacity and stubbornness accomplished one of the great breakthroughs in medical history.

Cataracts were known in Bible times. They dim the vision of people in advancing years. They obscure the lens of the eye, preventing adequate light from reaching the retina – the tissue lining the back of the eye that receives and transmits visual signals via the optic nerve. Some cataracts develop over a long time; others in a few months. They result in hazy, blurred or double vision; colors that may fade or seem yellow or brown. Vision is distorted, then lost.

Before Ridley, surgical intervention to deal with cataracts was extremely crude. One of the great "advances" in ophthalmology was when a Frenchman, Jacques Daviel, used a barber's razor and – without anesthetics – cut out the obstructing film from a man's eye in hopes of improving vision. The next advance was when a German, Heinrich von Helmholtz, in 1852 invented the ophthalmoscope. This instrument (still in use) beams powerful light into the eye and illuminates the retina. It nudged forward the techniques of surgery. But another 60 years elapsed before that marvelous instrument was fully employed.

In 1911, the Nobel Prize in Medicine was awarded to Alvar Gullstrand, of Sweden, who mapped the anatomy and mechanisms of the optical system – and more precise surgery became possible. A clumsy expedient was then devised: thick-lensed eyeglasses that provided substitute vision. People wore them, but not very comfortably. Distances were baffling and care had to be exercised watching where to plant one's feet. (The writer's elderly mother was so distressed by this handicap that she refused to have a second cataract-removal operation when the need arose.)

In those "bad old days," the post-surgical patient had to lie in a hospital bed absolutely immobile, for two weeks or more. Sandbags on each side of the head prevented movement, even while at sleep.

Status quo, decades drifted by again, until one morning in 1948. In the operating room of London's St. Thomas Hospital, the surgeon, Harold Ridley, was performing the Daviel procedure to remove a cataract. His assistant, not having previously observed this operation, innocently asked, "Are you going to put a new lens in to replace the absent part of the eye?"

Ridley says hearing that question was like a key turning in a lock: why <u>not</u> put a new lens into the eye? It had never been done – but why not try? If the eye's optical system could thereby be "repaired," might not sight of natural quality be restored?

He restudied Gullstrand's anatomy of the inner eye over many months. His conviction grew that – though audacious – the theory seemed sound. His next step was to consult peers. Unanimously, they scoffed. Even his father, an eye-specialist, strongly advised him not to try. But from his mother's side of the family, Ridley claims inheritance of a stubborn impatience with status quo. "They were very independent, very practical and inventive," he recalls fondly. A motto hung on their dining room wall: "Do it – and it's done." That decided him. Ridley carefully explained his theory to two patients. He told them the procedure was unprecedented, untested, and risked the possible loss of the eye.

"Things were different in those days. Londoners had come through the (World War II) blitz by helping each other when their homes were bombed out, when neighbors were killed and injured. There was a spirit of cooperation. There was trust. My patients knew we would do the best for them that we could – just as we had during the war. They volunteered to help advance the course of science. Had it not been so, my invention of the intra-ocular lens implant would not have happened when it did."

The odds against success were stupendous. "We needed what didn't exist – a chemically-stable, accurately-balanced plastic, lighter than glass, and so pure that it would not cause chemical damage to the inner eye. The refraction formula was unknown, as it required calculation of bi-convex curvatures to compensate for an aqueous environment. Lastly, we didn't have the adhesive necessary to anchor the new lens inside the eye."

Those were the medical mysteries. There were severe practical problems as well. "St. Thomas Hospital was directly across the Thames River from the Houses of Parliament and Big Ben. The entire area had been repeatedly bombed during the war. We worked in a half-ruined building with a makeshift lighting system. Our instruments were of 19th-century vintage – primitive. So were our medications. There were no anti-inflammatory steroids."

Nevertheless, on the 29th of November, 1949, the first artificial lens was inserted by Ridley into the eye of one of his volunteers. Would the foreign body inside the eye be rejected? Would infection endanger the eye? Even if these likely perils did not occur, would the implant restore normal vision?

It did. Fifteen more operations were performed on volunteers over the next nineteen months. "Sadly, some failed, mainly because the implants dislocated, lacking full support inside the eye. But not one eye was lost! Our rate of failure was no higher than with conventional cataract removals." They learned from failures even as they were developing surgical cogency from successes.

Secrecy was maintained until July, 1951, when Ridley reported his experiments and results at the Oxford Congress of Ophthalmology. The sensational news, headlined in professional journals everywhere, was "greeted with disbelief and often with hostility," he reminisces. His blue eyes flash and his cheeks flush. "I got furious reactions from some of the most eminent professors in the field – on both sides of the ocean:"

"This operation should never be done."

"It offends the first principle of ophthalmic surgery and could cause malignant disease."

"As long as I remain in charge of this department, no implant will ever be done."

"Dr. Ridley ... why don't you GO HOME?"

Instead, he persisted, and so did resistance within the profession. Older men, proficient in the traditional techniques, were satisfied with status quo – or were apprehensive about learning new ways. It is a familiar problem, Ridley says: "The greatest fear known to man is a new idea." Foot-draggers always oppose progress. (After the First World War, American Air Corps brass ridiculed Billy Mitchell's pleas to develop air power.)

Ridley's success did not inhibit critics from deriding him. They charged that he had plunged ahead too soon, that he lacked the technology for such delicate surgery. "It was a unique moment," he counters. "War survivors were public-spirited and eager to cooperate. Nobody then would have thought of demanding money if an operation failed. Had we not seized the opportunity, decades might have lapsed before anybody had the courage to try!" Quietly he added, "Maybe not even yet ..."

Slowly, slowly, the good news spread. Younger men – not hidebound by tradition and orthodoxy – experimented with the Ridley procedure. By 1967, a total of fifteen surgeons on both sides of the Atlantic were using it.

Fifteen – after fifteen years!

Why the reluctance to embrace a new technology even after it has been proved to be a great improvement? For the same reason sawyers in England destroyed the first water-powered sawmills built there. The first spinning jennies and sewing machines were wrecked by mobs. Only 100 years ago, coal miners tried to destroy a newly-built labor-saving machine designed for the pits – and were only deterred by sight of the inventor's gun.

Cases such as these are infinite in number, usually goaded by fear of unemployment. More difficult to understand are "professionals" who stand in the way of progress. When most of the world's ophthalmologists rejected Ridley's invention, he didn't lose heart, but persisted. He honed his techniques and narrowed his practice to only intra-ocular lens implants. Clinical experience and practice gradually produced improvements in the quality of the adhesive, the delicacy of surgical instruments, the precision of grinding and stability of the acrylic used to create lenses. Ridley preached his gospel wherever an audience could be gathered and a willing patient could be provided. With a small support team that included his wife, Elizabeth, he traveled the world ("We slept in palaces and mud huts!") to guide surgeons through their first transplants.

Imagine the discouragement he must often have felt, as year after year – for 35 years! – the laborious missionary work continued. Until a fateful day in 1986, when Ridley met Dr. David J. Apple, at an ophthalmologic conference in Israel. An American, Dr. Apple was one of the few who had acquired experience working with intra-ocular lenses and implants – even devised refinements. He was amazed to learn that Ridley was still not acknowledged by most eye specialists. Upon return to the U.S., he undertook a campaign of education in professional publications. It resulted in the conferring of an honorary doctorate on Ridley at the Medical University of South Carolina, in 1989. After 40 years of "laboring in the vineyard," that was his first academic honor! Since then, he has received virtually all the world's most prestigious awards in his field. He glows when he describes a medical convention in San Francisco, where, upon entering the huge hall, he was greeted by a standing ovation and presented with a tribute signed by 4,000 ophthalmologists.

Only six scientists before Ridley had been honored by the King of Sweden with a gold medal in Alvar Gullstrand's memory. He was appointed a consultant to Britain's Ministry of Defense and to the World Health Organization.

But Ridley quickly diverts further talk of personal satisfactions by praising "the Nightingales," the nursing sisters with whom he had worked at St. Thomas Hospital. "The Florence Nightingale School of Nursing" had been formed there, named after the woman who pioneered the profession in the Crimean War. In 1910, Harold had been dangled on the knee of the famous lady herself. He was three, she was 90.

Born into a family with strong traditions of service to others, he believes in the "give-back" principle. In 1967, he deposited all the money he had received from his parents' estates into a charitable foundation, and added to it thereafter, from his own earnings. The foundation helps to support aged ex-nurses, "some of whom were paid only fifty pounds sterling (now worth about $80) a year during their working lives."

Ridley stresses that his altruism was not unique. The makers of the plastic lenses and the technicians who ground them "charged me only cost price, less than one pound sterling (under $2) for the early implants." (But he admits that "firms who came in years later made vast fortunes from our work.")

Ridley's M.D. is from Cambridge; he also holds two Ph.D. degrees. But he insists on being addressed as MISTER. It is an old English tradition dating back to the era when only clergymen were healers – but were not permitted, for religious reasons, to draw blood. Surgery was performed by barbers. His grandfather, a minister and a school headmaster, was known as Dr. Ridley; but his father, an ophthalmologist, was addressed as "mister."

Harold Ridley entered medical practice at St. Thomas Hospital, London, 1932 – "the lowest point of the great depression," he notes. His first salary was under $400 a year. Ten years later, during the war, he worked in West Africa, caring for lepers afflicted "by worms so tiny as to be almost invisible. Ninety percent of the people had that disease; ten percent were totally blind." During World War II, he operated on many R.A.F. airmen who had suffered eye injuries from slivers of shattered cockpit Plexiglas when their planes were riddled by enemy fire.

He learned that plastic was chemically stable, non-allergic, and would be non-infectious. From that knowledge was born his career as a "revolutionary" – unafraid to depart from traditional teachings. "Youth is necessary for such adventure."

But Ridley cautions that young scientists should not "swim against the tide" simply for the pleasure of being non-conformists. Rather, look for the right opportunity, he says, be sure the goal is meritorious. Be not deterred by opposition. "Things don't just happen – they must be made to happen – with resourcefulness, creativity, and careful judgment. Nothing worthwhile can be accomplished overnight. It takes time to reach goals." Plus, he insists, tenacity and hard work. Opposition, frustration, and disappointments can be expected.

Hidden by a thick stand of tall trees and invisible from the road, the Ridley's modest "Keeper's Cottage" faces a swift-flowing river 10 miles outside Salisbury. Ducks float serenely downstream, carried by the current. Birds are constant visitors, fluttering in and out of feeders outside a huge window. Massive hand-hewn beams support the ancient ceiling. (I moved about in a semi-crouch throughout the visit.) It has been the Ridley's only home since his first years as a physician in London. They never felt the need to "move up."

At 88, Ridley looked 20 years younger, a full head of white hair crowning his ruddy complexion. Age slowed, but did not stop him. He learned to use the computer after his 80th birthday. Professional papers authored by him number nearly 100 as this is written. He wore a hearing aid, and had intra-ocular lenses in both eyes. Yet, he still rose at 5 a.m. every day to resume study of problems in ophthalmology.

To mavericks in any field of endeavor, he urges courage and independence. "If you have strong reasons to believe in your ideas, have confidence – face the brickbats and go ahead!"

To "authority figures," his advice is "give new ideas a fair hearing. Even if not perfect at first, there may be merit in them that could be developed with encouragement."

IGNATZ SEMMELWEIS
Reviled in life,
Revered in death.

In the hours before a woman expects birth of her baby, some fear pain; some prolonged labor; some birth-injury; some abnormality in the infant. Not so long ago, they feared worse – dying themselves. At the dawn of the twentieth century, 20% of women in labor died.

The famous American physician Oliver Wendell Holmes suspected that medical personnel themselves were carriers of "childbirth fever" during the normal course of gynecology examinations. He said so in print. He was promptly ridiculed by Professor Charles Meigs, an eminence among obstetricians: "I prefer to attribute (deaths in childbirth) to accidents or Providence, of which I can form a conception, rather than to a contagion of which I cannot form any clear idea." They battled – in the medical press – to "no decision" in 1843.

Decades later, in March 1879 – well after Louis Pasteur had published his discoveries about bacteria – a famous French physician could still insist, to the Paris Academy of Medicine, that the "miasm of puerperal (childbirth) fever could not possibly be caused by micro-organisms" (germs as they were then called). This arrogance so irritated the great Pasteur, himself not a physician, that he jumped to his feet and shouted: "It is the doctor and his staff who carry the microbe from a sick woman to a healthy woman!"

Obstinately, obstetricians continued to blame the fatalities on causes such as nutritional deficiency in the mother's milk; or "lochian suppression;" or "gastric-bilious disturbance;" or poor ventilation or poor food; or peritonitis; or "erysipelas of the bowels;" or even, said one authority, death followed a women's embarrassment at exposing her female genitalia to the sight of a man.

These surmises are quotations from German-language medical literature. In the 19th century, the German-speaking states of central Europe were in the vanguard of scientific research (and continued to be until Hitlerism). Fluency in the German language was vital to any practitioner, anywhere in the world, who hoped to keep abreast of findings in his specialty.

Far into the 20th century, "general practitioners" delivered babies and practiced surgery. Up to 30% of new mothers in England were maimed in childbirth or internally deformed – a startling fact discovered by Steve Humphries and Pamela Gordon in researching their book "Labour of Love." They also inform us that about 25,000 babies died annually during birth or within four weeks thereafter. Oddest of all their revelations is that <u>higher</u> rates of mortality during childbirth were reported among middle-class women than poor or working-class women – in some cities, nearly double! The reason? Ladies who could afford the cost, gave birth in hospitals – where they could pay for the "best" medical care; those who couldn't had their babies at home, attended by neighbor-women or a midwife.

Superior hospitals everywhere were (and are) "teaching hospitals." To teach surgery, they bought from Paris artificial corpses that were anatomically correct – containing 1,244 parts, each removable and re-usable. But Robert Knox, in Edinburgh, and Sir Astley Cooper, in London, declined to use this expedient. Cooper said "the practice of the most sensible and the most expert surgeons ... has been to visit the receptacles for the dead (mortuaries) for the purpose of (first) performing the operations which they were about to execute on the living ... Anatomy is our polar star." Progressive medicos in teaching hospitals on the continent adopted that policy.

In Vienna, Johann Klein, newly-appointed Chief of the Maternity Division in the University of Vienna General Hospital, was determined to keep his department abreast of the newest ideas and techniques in his specialty. Klein instituted the requirement that all doctors on his service perform pathology dissections every day.

Whereupon, the death rate of new mothers and their newborns in Klein's Ward One climbed steadily, until it exceeded – by more than six times! – that of Ward Two, where midwives were in attendance, not certified physicians. And then the death rate kept rising. Between October 1841 and May of 1843, 5,139 women entered Ward One. 829 did not leave alive – 16%. Many infants died, too. Many women approaching term induced labor at home to deliver before they reached the hospital, fearing assignment to Ward One.

The situation shocked Ignatz Semmelweis, a young Hungarian "resident" physician who had studied obstetrics at the Vienna Medical School, one of the first in the world to combine laboratory practice and bedside medicine. He then took graduate work in midwifery and was awarded a master's degree in that specialty. Even then, he didn't feel he was properly prepared as a physician, so he took another 18 months of intensive training in surgery, diagnostic and statistical methods. During this time, he performed dissections of all cadavers from the gynecology ward.

Determined to find the cause of the dread pestilence, Semmelweis set himself a tough regimen. Before his early-morning rounds, "I regularly examined obstetric cadavers in the anatomical theater ... I myself made the autopsy in order to compare my observations with ... (death certificate) findings." In this painstaking manner, he discovered a startling fact: the pathology reports on the mothers and on their newborn babies were identical. Both died – almost simultaneously – of the same disease.

Another day, an even more dramatic fact presented itself. Semmelweis was in a group of solemn men in street clothes that surrounded the dissecting table in the mortuary. Some wore suits with dried blood stains. A few had on blood-stained smocks. That was normal dress in surgery, in maternity – and in the morgue. Semmelweis had come to bid farewell to the corpse of a professor whom he'd deeply admired in medical school.

He smelled and saw something strange. A cadaveric odor emanated from the hands of the dissectors – the smell of poison. And he saw sores on the corpse – the same kind of sores he had seen that very morning – on the bodies of women who had died after childbirth. "There was forced upon my mind with irresistible clarity ... (that) I had seen so many ... (women) die from the same disease."

Acting on a hunch, Semmelweis joined the physician group next morning that would attend Ward One patients. Their duty roster had them, first, performing dissections in the morgue. They then moved to the maternity floor. The very same men who had been cutting up putrefying corpses, moments later performed internal examinations on women in labor. They did not wash their hands enroute.

Without asking permission from his Chief, Dr. Klein, Semmelweis issued orders that all doctors must wash their hands in a chlorinated lime solution before they examined pregnant women or women in labor. His mandate was grudgingly obeyed by the staff, assuming that the orders had been issued by higher authority. Within two months, the mortality rate in both wards had fallen to 2.3 percent.

He expected praise when he reported to Klein: "Puerperal fever is caused ... by the hands of the physician who examines the childbed patients ... it is of utmost importance that he should clean his hands properly before a visit."

Not praise but anger was his reward. Klein bitterly resented being told his business by a young man. The Hungarian hick was presuming to contradict traditional procedures that senior doctors had long been practicing!

History stepped in and prevented a confrontation between the young man and his superior. The social revolutions of 1848 swept across Europe, a class-struggle by working people against the tyranny of feudal autocracies. In Vienna, intellectuals and professional men formed an "Academic Legion" to join the revolt, as they did during the Spanish Civil War in the 1930s. Semmelweis was an idealist of that stripe, and – along with most of the young doctors – enlisted.

The revolutions were quickly crushed in every capitol city by weight of arms. Thousands were killed. Semmelweis was in one of the last groups to surrender a stronghold. He escaped unharmed and returned to the hospital.

During the chaotic weeks of street battles, few doctors were on duty, so few autopsies were performed. Not a single death was reported during that period – woman or baby! It became obvious that Semmelweis was right: the infection was being transmitted on the hands of medical men.

The senior official at Vienna General Hospital invited the young doctor to address the city's Association of Physicians on May 15, 1850. The most prominent figures in the profession attended – and applauded. Klein, however, sat silent. The youth's "insult' in going over his head made of him a lifelong enemy. His chance for revenge soon arrived. Semmelweis' contract appointment on the hospital staff expired. Extensions were usually automatic. Every one of Semmelweis' superiors endorsed reappointment and promotion to full professor. Klein intervened with the Vice-Chancellor of the University, who blackballed the appointment.

Semmelweis was out of work for a year and a half. Finally, a job was offered – but with humiliating conditions. He was prohibited from teaching dissection on actual bodies; he was ordered to demonstrate on dressmakers' dummies only! He was also prohibited from granting permission-to-practice certificates.

Though still in his twenties, Semmelweis was spirited; he packed his bags and went back to Budapest, twelve years after he'd left. By then, both his parents had died. Two of his brothers were fugitives, having fought in the 1848 revolution. Secret police and informers were everywhere. The Academy of Science was shut; the medical society could only meet when a policeman was present.

Nevertheless, the Semmelweis reputation had traveled. He was appointed director of obstetrics at the St. Rochus Hospital in May 1851, and continued there until 1857. He introduced chlorine disinfection, and the scourge of childbirth fever was stamped out. (Death rate from puerperal infection sank to under 1%.) His fame spread and his private practice flourished.

Thus passed five happy years. In 1855, Ignatz Semmelweis was also appointed professor of Theoretical and Practical Midwifery at the University.

Resistance to new ideas is not uncommon; rather, it is the norm. Despite proofs as irrefutable as Semmelweis' statistics that a chlorine rinse would save numberless lives, the simple prophylactic procedure was elsewhere still widely ignored.

During this period of status quo – it lasted for years – doctors-in-training were not properly educated to take sanitary precautions before touching a pregnant patient. Thousands of young women died. Their infants, too. In one German university hospital, a student testified that not one new mother survived after giving birth. "All died from puerperal fever," he said.

"My doctrine," Semmelweis demanded, " ... (should) be disseminated by teachers of midwifery, until all who practice medicine, down to the last village doctor and the last village midwife, may act according to its principle ... banish the terror from the lying-in hospitals ... preserve the wife ... the mother ..." He expected that his carefully-detailed cause, effect, and prescription would become a standard text in medical schools.

But Semmelweis had been overwhelmed with work during the 10 years after his return to Budapest, serving on hospital committees, library organizations, and seminars on clinical procedures. So it was not until 1861 that he published, in book form, "The Etiology, Nature and Prophylaxis of Puerperal Fever," recounting the historical background of his great discovery.

Amazingly, many reviews of the book were unfavorable! In medical journals, some reviews scoffed at the Hungarian's "solution" to the long-familiar, but still mysterious disease. Few remembered Oliver Wendell Holmes' warning: "Whatever indulgence may be granted to those who have heretofore been the ignorant causes of so much misery, the time has come when the existence of a private pestilence ... should be looked upon not as a misfortune, but a crime."

What so agitated Holmes, haunted Semmelweis. He became embittered, irascible. He wrote blistering denunciations to his detractors (later published): "... Your teaching is based on the dead bodies of ... women slaughtered through ignorance ... I will put an end to this murderous work ... If you ... continue to teach your students ... (a wrong) doctrine ... I denounce you before God and the world as a murderer ... a medical Nero." This intemperate language was unprecedented at that time in the stiff, formal world of academe. (It still is offensive! In discussing Semmelweis with my professor at Cambridge University, I learned some historians – he included – believed the Hungarian was at fault to defame other doctors so bluntly!)

He didn't spare himself: "God only knows the number of women whom I have consigned prematurely to the grave ... if the misfortune is not to remain permanent, the truth must be brought home.

He became infamous in medical circles as "the Hungarian crank." Frustration turned him into a fanatic. Then a neurotic; finally, a psychotic. He stopped strangers on the street and harangued them about puerperal fever. Soon he slipped beyond the edge. On the 31st day of July 1865, family and friends, wife and infant child, accompanied him on his last journey – to an insane asylum in Vienna. He died there.

In the city of his birth, Semmelweis' mausoleum is crowned by a life-size statue of him; a young mother is seated at his feet, cradling her infant. She looks gratefully upward, representing the millions of lives this man saved since his teaching was finally accepted.

NIKOLA TESLA
Once more famous
Than Tom Edison

He could have become the richest man in the world, but Nikola Tesla died alone and broke. "The common man loses one of his best friends," said the Vice-President of the United States, Henry Wallace, at the funeral in New York's Cathedral of St. John the Divine. Three Nobel Prize laureates delivered eulogies, honoring "one of the outstanding intellects of the world who paved the way for many of the technological developments in modern times."

Croation-born Tesla's amazing odyssey had started in 1886 when he arrived in New York. That very day, he went to the office of Thomas Edison, for whose company he'd worked in Paris. It was not a good day for Edison – "wizard of Menlo Park." Since early that morning, he had been toiling to repair a trivial short-circuit in the lighting system at Mrs. Vanderbilt's mansion. There had been a small fire in the wiring, but it could lead to large consequences. If the "queen of New York society" weren't quickly placated, the very vocal lady could cripple Edison's infant lighting business.

He fixed the problem – after many hours' grimy work. But when he got back to his office, an even bigger problem had to be faced. The Edison electrical system on an ocean liner, which was docked in the East River, was dead. Over the telephone, the captain made clear that delay of the sailing schedule would cost the shipping company heavily – and Edison would be held responsible.

As Edison hung up, Nikola Tesla barged in – with no appointment. In awkward English, he sputtered that he had invented a new type of motor. It used a rotating magnetic field to generate "alternating-current" electricity by induction.

It sounded like gibberish to Edison, who growled that alternating current didn't exist. He had always worked with direct-current electricity. His success was based on inventions using DC. At 39, he had never heard of rotating magnetic fields, nor of induction. He sputtered "Spare me that nonsense! We're set up for direct current in America!"

But the crazy foreigner apparently knew something about electricity, and the ocean-liner crisis could be very costly. All his senior technicians were on other assignments. So Edison offered him a job – if he could fix the equipment on the ship. That day.

Nikola Tesla rushed to the dock. He worked through the night, repairing the badly-worn Edison machines. By dawn, he had restored power so the ship was able to sail. Edison was delighted, and hired the man at $18 a week, an entry-level wage in those days.

Tesla became a trouble-shooter for the Edison Company, going from emergency to emergency, patching and re-wiring short-circuits and replacing worn-out equipment. He told Edison how all his motors could be rebuilt to function more reliably; the savings could be enormous. "Get started!" Edison said. If the anticipated economies were in fact realized, he promised a $50,000 bonus!

Tesla worked day and night for many months. he not only rebuilt old Edison machines, he invented automatic controls that greatly reduced wear and tear. Patents were taken out – in Edison's name.

Tesla didn't mind. He was working for that promised $50,000 bonus. When he asked for it, Edison laughed in his face. "Tesla, you don't understand American humor!" He offered instead a $10 weekly raise.

Tesla didn't think it was all that funny and quit immediately. (Edison later admitted losing him was the worst mistake of his life.)

Tesla was under 30. Earning $18 a week, his savings weren't much. "Intangible assets," however, were enormous: self-confidence and invaluable "hands-on" experience – building upon a lifetime of theoretical as well as practical work.

As a five-year-old in rural Croatia, he had built a water-wheel – without paddles – to harness energy from the stream that flowed through the family farm. He took clocks apart to understand how they worked, then put them together again. He asked questions: when people walked on snowy country roads at night, why did they create a luminous trail? When he petted the family cat, why was a "halo" formed? This strange power in the air worked magic!

His father, Milutin Tesla, was a minister in the local Serbian Orthodox Church. He led his family in readings from the Bible every night; and from literary classics and poetry. Aware of Nikola's exceptional intelligence, he put him on a training program to sharpen memory and stimulate imagination. He assigned mathematical problems for the boy to solve – in his head, without pencil or paper. When adults conversed, Nikola was expected to memorize what each person said; take special note of their grammatical errors; then repeat their words and correct their errors.

In bed at night, the boy conceived of mechanical devices to save time and labor on the farm. He designed every part of each contraption in his head. When he entered primary school, he learned basic English, Italian, German and French. At ten, he was introduced to physics and the principles of electricity.

By 1875, his knowledge of natural science was so advanced, he was awarded a tuition-paid fellowship to a technical institute in Austria, the governing power over Serbia. In his unheated dormitory room, he rose every morning at 3 a.m. to study. In class, he was fascinated by a machine using direct-current electricity that was operable as a motor and also reversible, to become a dynamo. But why did energy thus generated lose strength in transmission? Why was that typical of all DC equipment?

Answers were not forthcoming in the classroom. He "dropped out" to speed the learning process at work. With the money he earned and saved, he enrolled at the University of Prague in September 1878. When his father died suddenly, his widowed mother needed his help. His formal schooling ended and the boy became the family's main support.

In Budapest, Hungary, a central exchange was being built for a new fangled American invention, the telephone. The job paid well, and left him with spare time. On long walks, he chewed on the problem of power-loss in transmission from a direct-current motor. In a Budapest park one day, the solution suddenly came to him. "Like a flash of lightning," he said years later. "In an instant, I saw it all, and, with a stick, drew in the sand the diagrams which were illustrated in my fundamental patents of May, 1888 ..." A set of electromagnets was to be activated in sequence by multi-phase current. The influence of the phased currents on each other should cause the magnetic field to spin and become, by induction, an electric motor. It is a simple concept – but nobody before him had thought of it. (Induction motors are now commonly used to drive most field and factory machines.)

Next morning, he described the idea to his bosses. It was so different from any electrical technology then known, they laughed at him. Faced by closed minds, he quit, and used his savings to go to Paris. There he got a job with the Continental Edison Company. For two years, he worked as a trouble-shooter, solving problems for Edison customers all over Europe. He also built for himself a small dynamo to change mechanical energy to electrical energy. It worked exactly as he had theorized – and reversed direction on command. Once again, he tried to interest management in looking at his model. Once again, they shook their heads. He decided to go to America, where new ideas in technology were flourishing. The very day he arrived in New York, he went to Edison's office. In his pocket, four pennies. In his head, knowledge of 12 languages. Under his arm, a book of poetry and a roll of mechanical drawings.

But Edison, with fixed ideas, had become obtuse, and Tesla, once again, was out on the street. Literally out on the street. It was 1887. In little more than a century, the country had fought three wars, and was in an economic recession. Tesla could not find work for which he was qualified. So he became a ditch-digger at two dollars a day. The crew foreman soon recognized that the quiet foreigner, reared on farm labor, was his hardest worker. They talked during lunch "breaks." When he learned about Tesla's experience, he introduced him to a friend who was an engineer.

The last decades of the 19th century were feverish with ideas and inventions. Venturesome investors were making fortunes. The new friend assembled capital to organize a company so Tesla could develop his revolutionary alternating-current concept. He went to work on a prototype.

It took only six months to build two small AC induction motors and send them to Washington. Patents were granted. He made dynamos to generate current and transformers to reduce voltage, and automatic controls. Thirty additional patents were awarded.

All this came to the attention of George Westinghouse, founder of the giant company that we know today. He quickly grasped the theory. Most enticing was the promise that alternating current can be transmitted over great distances from a single generating plant, whereas, because of Ohm's Law of resistance, Edison needed 121 central power stations to serve just New York City with DC electricity.

Westinghouse came from Pittsburgh to meet Tesla. Each perceived the other's intelligence and integrity. They quickly made a deal. One year after he was earning his bread as a laborer in the streets of New York, Tesla was paid $60,000 for exclusive use of his patents by the Westinghouse Company.

Plus a royalty – and what a royalty! For every horsepower-unit of electricity that would be generated from Tesla-designed equipment, he would receive $2.50.

The industry was in its infancy. Every city, every factory, farm and home in the world would have to be wired. Unlike other energy resources – oil, coal, forestry products – electricity would not suffer depletion. The more wires that were strung, the more markets would be created for electrical appliances and devices, requiring ever more electricity. The demand would be limitless.

Edison was famous and popular: his many inventions were already in common use. He recognized his monopoly was in danger, and fired the first salvos in open war: "Westinghouse will kill a customer within six months after he puts in a system of any size." Newspapers printed his prediction.

In those years, there were few regulations governing business ethics. History books refer to it as the "era of the robber barons." Tesla's friend Mark Twain, described their ruling passion: "Get money, lots of money. Get it quickly ... Get it dishonestly if you can, honestly if you must."

Edison hired an electrical engineer named H. P. Brown, who devised a smear campaign to discredit Westinghouse and Tesla. During the summer of 1888, he paid children 25 cents for every cat and dog they stole and brought to him. At public exhibitions, he shocked the animals to death, one by one, with 1,000 volts of AC power. Journalists reported the evidence of their own eyes – alternating-current was lethal!

In city after city, town after town, Brown electrocuted small and large animals, reminding viewers that people who used alternating-current could also be "Westinghoused."

George Westinghouse, though, was no pushover. He produced testimonials from customers to prove that alternating-current was safe – and far more economical. AC motors require fewer parts, hence were cheaper to build, smaller and lighter per horsepower, and much more durable. These were provable facts. So Westinghouse won the contract to light the 1893 World Fair in Chicago, providing Tesla-designed engines, dynamos and alternating-current generators. Then he was awarded another history-making contract in 1895 – to build the first hydroelectric power system in the world, and light the city of Buffalo with energy generated by Niagara Falls.

The struggle between Edison and Westinghouse went on for years. Like every war, it cost both sides heavily, depressed their profits and their share value on the New York Stock Exchange. Edison was the first to admit defeat when he faced insolvency. He was absorbed by a newly-organized J. P. Morgan "conglomerate," the General Electric Company. Morgan was using the same technique with which he'd swallowed many thinly-financed railroads and steel mills in the early days of those burgeoning industries.

Having ingested Edison, the rapacious financier went after Westinghouse. The "blitz" was described in <u>FRENZIED FINANCE</u>, in which the author, Thomas Lawson, painted a graphic picture: "From all the stock market sub-cellars and rat holes ... crept those wriggling, slimy snakes of bastard rumors ... (claiming that) Westinghouse had mismanaged his companies ... (and was) beyond extricating unless by consolidating with General Electric."

The Morgan octopus saw to it that the banks withheld working capital, crippling Westinghouse production. The industrialist knew the constriction would tighten. He sought to merge with several smaller electrical manufacturers that were still modestly profitable — and had not yet been swallowed by GE. Combined, their total assets and operating statements would be sufficient to warrant commercial financing. The new company could thus withstand the Morgan siege. It would still carry the Westinghouse name, and he would continue as its chief executive.

The partners-to-be made one condition ...

The contract with Nikola Tesla was unacceptable to them. The flow of royalty-money to him diluted profits. Millions were already owed him. Westinghouse was told he had to get rid of the deal and wipe out the debt — or the reorganization and salvage of his company would not go ahead. To avoid Edison's disaster, he went to Tesla, explained the crisis.

Tesla asked the one question that was more important to him than money: "If I give up the contract, you ... retain control ... (and) proceed with your plans to give my polyphase system to the world?"

Westinghouse responded "Your system is the greatest discovery in the field of electricity ... I intend to ... put the country on an alternating-current basis."

According to Margaret Cheney, Tesla's biographer, Tesla then said: "Mr. Westinghouse, you ... believed in me when others had no faith ... to save your company so that you can develop my inventions ... I will tear the contracts to pieces, and you will no longer have any troubles from my royalties."

He did so. The reorganized Westinghouse Company paid Tesla $216,600 to wipe off the outstanding debt and acquire exclusive rights to all his patents without payment of further royalties.

Did he make a bad deal?

He could have refused George Westinghouse's appeal. He knew that AC motors based on his patents inevitably would have been manufactured; within 30 years, over $50 billion went into building induction motors and power transmission systems, based on Tesla patents. His royalties would mount every year — to astronomic levels. He would have become the wealthiest man in the world.

But remember — he'd arrived in America with only a few pennies in his pocket. America gave him opportunity denied him elsewhere. In four short years, he became a "rich" man — the worth of a dollar 100 years ago was vastly greater than now.

He was free to be a true scientist. "I do not think there is any thrill," he said " ... like that felt by the inventor as he sees creation(s) of (his) brain unfolding to success." He had formidable tools — a lifelong study of physics, mathematics and chemistry. (Unlike Edison who focused on limited objectives — "practical" ideas likely to have quick cash-in prospects.) When the New York Times carried a front-page story November 6, 1915 that the Nobel Prize would be awarded to Tesla and Edison jointly, the report further stated that Tesla would decline to share the honor with a "mere tinkerer."

This story may be apocryphal — but likely. L. Sprague de Camp, in his fine book "The Heroic Age of American Invention," describes a paranoiac as one "whose ego is so inflated and so pathologically sensitive as to make him ... self-conceited, suspicious, truculent, vindictive, obsessed ... Tesla, for all his genius, is a case in point."

During the years after he shook hands with George Westinghouse, Tesla produced an extraordinary body of work. Among testimonials delivered on the occasion of his 75th birthday in 1931, he was hailed both as an "inventor and originator, the greatest ... in all history ... (his) revolutionary discoveries ... have no equal."

He was famous. His every pronouncement was seized upon by newspapers and magazines. His tall gaunt figure attracted attention wherever he walked. People dined at his hotel in order to see the genius folding a napkin, fresh for each course, from a tall pile. He caused a sensation in 1898 by announcing that he could operate a boat by remote control. He hired Madison Square Garden to prove the claim, and did so. For nearly a year from May 1899, he conducted experiments in wooded areas around Colorado Springs Colorado, and demonstrated that electricity could be conducted through the earth — without wires, and despite atmospheric disturbances and irregular terrain. He lighted 200 lamps from a distance of 25 miles! But these were carnival stunts, not money earners, and the Westinghouse payout was shrinking.

Upon return, he undertook construction on Long Island of a 145-foot tower with a perfect circle of copper wire at its top measuring 100 feet in diameter. He planned to use it for radio transmission abroad. Money was not available to complete that project, then World War I intervened. Tesla had been born in the Austro-Hungarian Empire, with which the U.S. was at war. Washington issued order to destroy the tower, fearing it would be used by spies to transmit espionage.

Thereafter, Tesla was almost always strapped for funds to realize his futuristic ideas. Often he had to stop work and scrounge for new capital. Impatient with details, Tesla's independence was costly to maintain. The Westinghouse money ran out eventually. He was so immersed in his work that he had no mind to collect royalties from patents.

His reputation was world-wide; some people invested very large sums. Others — employees, too — made personal loans. Tesla so feared J. P. Morgan that he refused an offer from him for major financing because it might entangle him in an Edison-type strangle-hold. Instead, he accepted a loan, surrendering as collateral 51% ownership of patents basic to the technology of radio.

Yes, it was his invention; Nobel Prize Laureate Guglielmo Marconi came later, "borrowed" ideas from where he could, and developed radio commercially. His patent, No. 7777, filed in 1900 was vacated by the U.S. Supreme Court in 1943, in favor of earlier Tesla patents. Tesla-patented equipment is still used today in every radio and television set.

Just a few of the other patents Tesla held beside the epochal invention of alternating-current transmission of electrical power: carbon-button lighting (20 times brighter than incandescent); the automobile speedometer; medical diathermy; vertical takeoff-and-landing aircraft (VTOL). An entire book is devoted to his patents — a small percentage of the inventions on which he did basic work. He conceived and initiated work (later completed by others) on cosmic-ray research that led to development of the cyclotron; artificial lighting for weather control; the electron microscope; biofeedback; earthquake early-detection instrumentation; solar heat; lighting for indoor photography; television; radar and sonar; guided missiles; a radio-controlled torpedo to locate and destroy enemy mines and ships; the electric clock; robotics; fluorescent lighting; high-voltage bombardment of iron for conversion into heat; AC/DC transformers; 110/220 AC converters. His two last inspirations are still being worked on by engineers: a "bladeless turbine" that would be 50% more efficient than those in use (then and now), greatly reducing the world's dependency upon petroleum; and wireless transmission of electrical energy from continent to continent, across oceans.

At the end, he lived in a cheap hotel in midtown New York. Almost the only furniture in his bare room was a large safe in which his notes on uncompleted projects were locked. He died at 86, January 7, 1943. His last few dollars were left to a friendly bellhop to buy food for his talisman, an all-white dove that came to his window every night.

THE WRIGHT BROTHERS
Started late
But won the race

>"Our ancestors ... looked enviously on the birds soaring freely through space, at full speed, above all obstacles, on the infinite highway of the air."

Wilbur Wright penned these lines. This is the story of why he and his brother, Orville, realized that ancient dream. Many before them had tried:

Armen Firman, an Arab savant in the 9th Century, tried to float down from a tower garbed in a voluminous cloak. He expected the garment would support him like a parachute. It didn't.

In the 11th Century, an English monk named Eilmer jumped with homemade wings from the top of Malmesbury Abbey. His landing was rough; both legs were broken and Eilmer was crippled for life.

Leonardo da Vinci proposed to use technology as an assist to flying. Perhaps more imaginative as an engineer than as a painter, he designed a propeller-driven helicopter, with a stabilizing tail assembly. He also envisioned a contraption that, strapped to a man's back, would enable him to float under wind-power. His then-patron didn't provide the means to realize Leonardo's futuristic ideas.

A few years later, John Damien launched himself from a tower on Stirling Castle in Scotland. Luckily, only his hip was broken.

During the golden age of Moorish creativity in Spain, a doctor named Abbas ibn-Firnas strapped himself into feathered wings that flapped when he moved his arms. He had a rough landing.

And on and on, in Italy, France, Germany.

Invention of the hang-glider in the 19th Century at last satisfied man's ambition to float in the sky. And greatly whetted his appetite for powered flight. Power to carry him, like birds, wherever and whenever he wishes to fly. As the 19th Century waned, that wish became an obsession. At the turn of the century, the obsession developed into a race to be first. This story is about the last lap in that race.

On a bleak December day, people gathered near Kitty Hawk, on the Outer Bank of North Carolina. One onlooker made note that "the wind was blowing at about 20 miles per hour, hard enough to polka-dot Albemarle Sound behind us with whitecaps." The onlookers surrounded a small, bi-wing hang-glider with a tiny engine. The pilot strapped himself in. Several men tugged the 'plane into motion, along the takeoff track laid on a sand-dune. The craft came rolling down that track, gathering speed, then lifted off and (as a witness wrote later) "then settled with a thud on the sand." Three times, the pilot's attempt to rise into the air succeeded — momentarily — before the craft again fell to earth.

The aviator was a man named Ken Kellett. The bi-wing aircraft was not the original Wright Flyer, but an exact replica. The occasion was the 75th anniversary celebration of Orville Wright's first successful flight December 17, 1903. An audience of 3,000 witnessed Kellett's failed attempt in 1978, whereas only about a dozen had witnessed Orville' success.

On that occasion, one of the onlookers, a boy, raced to the one-room post office at Kitty Hawk and burst in, shouting "They have done it! Damned if they ain't flew!" It was a surprise, a sensation, the world over, in 1903. Many men before Wilbur and Orville Wright — better educated, seemingly better qualified — had tried and failed. Hiram Maxim had boasted he would "build a flying machine that would lift itself from the ground." He had more than enough money to build it. He was chief engineer for a pioneer electric utility before inventing, in 1884, a "machine gun" capable of firing 600 bullets per minute. (The U.S. Army and Navy turned down his invention; but the British were more far-sighted. They bought the rights to his patent. With his assumption of English citizenship, he was granted a knighthood. Royalties made him rich.)

He was fascinated with the idea of flying; studied wing forms and propellers for an airplane. He designed a lightweight steam-powered engine that — he projected — could lift a craft off the ground. With such a motor, he said, "we will very soon give you a successful flying machine."

On a rented country estate in southern England, he built the body of a gigantic airplane, eight feet wide. The craft had a long 40-foot "fuselage" whereon he mounted two 180-hp steam engines with pusher-propellers 17 feet long. The "power plant" assembly included a fuel tank and boiler. The 'plane had a 107-foot wingspan with a total lifting surface of 4,000 square feet. It weighed 8,000 pounds.

Four huge wheels rested on steel tracks; two parallel smaller ones were mounted overhead above the plane so that the craft, when airborne, would be restrained from rising more than a few feet. In an 1893 test-run, Sir Hiram climbed aboard, started his engines and pushed boiler pressure to its limit. The anchor ropes were released and the craft, according to English journalist H.J.W. Dam,

> "shot forward like a railway train ... big wheels whirring ... steam whistling ... waste pipes puffing and gurgling, (it) flew over 1800 feet of tracks."

Tinker-time followed, making improvements on the wings and tuning the lightweight engines. Finally, July 31, 1894, Maxim prepared to repeat the test run. With the restraining rope released, he roared forward 600 feet along the steel rails, building to a speed of 42 mph. The ponderous airplane miraculously rose — until it touched the overhead rails, whereupon one snapped from the impact and sudden friction. The broken rail flew into the propeller. Maxim shut off the engines; had he not, the machine would have crashed — no provision had been made to control it during free flight. Maxim, having proved that an engine-driven, manned airplane could leave the ground, never flew again, nor permitted his airplane to be flown, nor continued any other aeronautic experiments.

If not Maxim, it was expected that the winner of the race toward manned flight would be the famous Otto Lilienthal in Germany. During the same years that Hiram Maxim was developing his project, engineer Lilienthal was concentrating on a different aspect of the challenge: how to control flight. In his famous book, "Birdflight as the Basis of Aviation," he had predicted that "The manner in which we have to meet the irregularities of the wind, when soaring in the air, can only be learned by being in the air itself." He was, thus, the opposite of Sir Hiram in his thinking.

Born in 1848, Lilienthal established a small factory in 1880, to build his gliders. He made dozens of experimental flights from nearby hills, strapped onto a bench that was suspended from the junction of two wings, their frame fabricated from flexible willow saplings over which fabric had been stretched. The wings were cambered, like a bird's, and actually could be folded. To control maneuvering, he shifted his weight from side to side, and leaned forward or backward. Thus he was able to bank, to rise and descend; his body in motion constantly as he soared to altitudes over 1,000 feet. This extraordinary skill attracted worldwide interest. In 1896, a reporter representing a Boston newspaper described a flight when Lilienthal, on a steep hill,

> "faced the wind ... Presently the breeze freshened; he took three rapid steps forward and was instantly lifted from the ground, sailing off nearly horizontally from the summit. He went over my head at a terrific pace, at an elevation of about 50 feet, the wind playing wild tunes on the tense cordage of the machine ..."

Another visitor from America was Samuel Langley, Secretary of the Smithsonian Institute in Washington. He came away with an essay written by Lilienthal, in which stress was made that the essence of gliding required "a constant and arbitrary correction of the position of the center of gravity."

Interpreted into laymen's terms, Lilienthal thus explained his reliance upon strenuous physical exercise to control his craft. Unfortunately Lilienthal didn't adequately assess the fatigue that might build from the constant, aerobic exercise. That omission proved fatal. On August 9, 1896, Lilienthal's glider was soaring buoyantly when suddenly the friendly wind turned ugly — his glider was struck by a powerful gust that hurled it skyward, into near-vertical position. Lilienthal desperately threw his weight forward, trying to bring the glider's nose down before stalling. It was too late. The glider hung, nearly vertical and motionless for a split second, then responded to the fateful pull of gravity. It flip-flopped, plummeting into a dive and crashed into the ground. Lilienthal died the next day of a broken spine.

That same year, Lilienthal's acolyte, a Scotsman named Percy Sinclair Pilcher, built a similar camber-wing hang-glider. Similar, but with an important design improvement — a hinged tail unit that could tilt the craft up and down in flight and thus combat the wind when it misbehaved. The next three years Pilcher devoted to designing a 4-horsepower gasoline engine. Then, needing more capital to continue work, he invited potential investors to an exhibition flight of the powered glider. The day before the scheduled performance, the craft was towed to the top of a hill and anchored there for morning takeoff.

Night moisture had impregnated the frame so that when Pilcher took off into the wind and the glider soared to 30 feet above the ground, a bamboo rod in the tail-assembly snapped and the weakened tail collapsed. The crippled craft dropped like a stone to earth. At 33, Percy Pilcher never regained consciousness.

If the race-winner was not to be an Englishman or a German, surely he would be a Frenchman. France had long been in the vanguard of experimentation. 56-year-old Clement Ader was the prime candidate. Seated in his "Avion" October 14, 1897 on a field near Versailles, France, he revved his two engines to a speed he estimated would be needed to become airborne. At the very last moment before he would command release of the restraining ropes, a gust of wind hit the craft and spun it around, damaging a wing. Ader switched off the engines and climbed down, prepared to make repairs. The 'plane had never left the ground; but the potential investors had seen enough. They lost interest in what was obviously an impractical project.

These discouragements were well known to Octave Chanute, America's most respected name in aeronautics — and the world's first serious historian of the new science. Brought to the U.S. as a small boy from his native land, France, he had been tutored in a variety of trades enroute to becoming an engineer. Prominent by mid-life in the new railroad industry, he became wealthy by counseling rail tycoons, and then by heading his own rail line. He turned to full-time study of aeronautics. In 1891, the first of 27 essays he wrote on that subject appeared in "Railroad and Engineering Journal." The articles were collected, augmented, and his soon-famous book "Progress in Flying Machines" was published in 1894.

He engaged young Augustus M. Herring to build a glider he had designed. By June 22, 1896, Herring was ready for the first trial on the south shore of Lake Michigan, where strong winds blew almost constantly. Beach sand had, over centuries, built up into high dunes. Herring made nearly 100 test-flights from those hills — achieving speeds up to 17 miles per hour at a height of over 100 feet. Data learned in the course of these experiences were synopsized by Chanute in the 1897 "Aeronautical Journal":

> "the operator was compelled to shift his weight constantly,
> like a tight-rope dancer without a pole, in order to (maintain a)
> ... center of gravity ... and to avoid being upset."

Chanute and Herring dabbled with design changes, adding wings — the glider at one point had 12 of them, each six feet long, three feet wide. As Herring's piloting experience grew, superfluous wings were dropped, one by one, until only two remained. Distance glides increased to nearly 400 feet, up to 30 miles per hour.

Herring then took the giant step of attempting powered flight. October ll, 1898, he tested the glider with a two-cylinder compressed-air engine. But, burdened with the added weight of the engine, it never gained altitude. Forward progress was only made on the same level or slightly downward from the launch point on the dune. The wind, aided by gravity — not the engine — lifted the craft off the sand; the landing was lower in altitude than its embarkation point. Herring claimed that he was the first pilot to fly skyward in a powered machine — but history doesn't honor flights that do not result in climbing higher than takeoff.

In his 1891 book, Chanute had made the strong point that aeronautics requires "the study of men of broad knowledge, and accurate training, and ... (is not) to be considered ... (a) hobby." It was singularly appropriate, therefore, that the transition from hobbyist to aviator was next attempted by a quintessential professional, Samuel Pierpont Langley, widely respected as a professor of astronomy and physics. He attended a Chanute seminar in 1886, at the American Association for the Advancement of Science convention in Buffalo, N.Y., and "caught the bug." As Secretary of the Smithsonian, Langley had the prestige, the money, facilities and staff to fulfill his prediction that "mechanical flight is possible with engines we now possess." Like Maxim, Langley believed that the key to success was power. Like Maxim, like Ader, he was less concerned with the ratio of power to weight; and of wing design to the vagaries of wind. Langley invested the following years into glider and engine design. Finally, his project reached its climax on a May day in 1897. Accompanied by his friend, Alexander Graham Bell, famous for his invention of the telephone, Langley went by train and ferry to Scott's Island on the Potomac, south of Washington. This was the headquarters of a private club for wealthy duck-hunters, which numbered among its members William McKinley, President of the United States. (He later authorized the Army to grant $50,000 for Langley's further development work, looking toward a soon-to-come war with Spain over Cuba.)

Langley brought to the launch site on Scott's Island two identical, small gliders. A scow, anchored in a cove, had a shed on its deck which at night housed the two models. Each was 10 feet long, each had two wings that spanned 14 feet, attached by struts and guy-wires to a steel-tubing fuselage. Each mounted, rear of the wings, a push-propeller and a one-horsepower steam-engine. Each weighed 26 pounds. Fastened to the roof of the shed was a horizontal 20-foot rail that projected 16 feet over the water.

The first airplane was hoisted onto the rail. A heavy spring held it in place, until Luther Reed, Langley's chief carpenter at the Smithsonian, was instructed to release the restraint. When he did, and the miniature plane bolted forward, one of the guy wires snapped. The craft shot out from the stern of the boat, and into the river, smashed into uselessness.

Later in the day, guy wires having been strengthened on the second glider, another try was made. The engine was stoked to 150 pounds of steam pressure; the craft was catapulted along the rails. Reaching the end, it lost altitude momentarily, then rose into the wind at a 10 degree angle. Held to ground-control by kite-type lines, it circled gracefully around the boat for a minute and a half at a speed over 20 mph until it was 100 feet above the water. Then, fuel exhausted after 3,000 feet of flight, the craft splashed down in the river. The men manipulating the control-lines subsequently succeeded in keeping the craft aloft nearly two minutes. Alexander Bell excitedly clicked the shutter of a camera throughout the flight.

The crucial next step — manned gliding — could now be undertaken by Langley. It would take another five years to reach that point.

The 20th Century was about to dawn; the stage was set for new heroes to appear. The Wright Brothers were unlikely candidates for starring roles. They were not engineers, had not been to college at all. They were not experienced glider pilots; they didn't have an equipped machine-shop, skilled employees or deep pockets. They were bicycle-repair mechanics. And geniuses.

Normally, little attention is paid to the birth and rearing of a genius. To understand the seeming-miracle wrought by the Wrights, however, justice demands attention to a few biographical details. Their father, Milton, bishop of an evangelical Protestant sect, had to travel throughout the Midwest, almost constantly. In Milton's frequent absences, the boys' mother, Susan, crafted toys; designed and sewed clothes; repaired household appliances; built sleds. Orville and Wilbur helped. The boys' older brother, Lorin, invented improvements in the hay-baling machine. The point cannot be too strongly stressed: the Wright family was familiar with using their brains and their own hands to achieve their goals.

In this environment of manual competence and "can-do" self-help, Orville and Wilbur built their own lathe, devised toys and gadgets. Wilbur was brilliant in trigonometry, and a keen competitor in sports at Central High School in

Dayton, Ohio. Orville never finished that school because he was more interested in building a job-printing business and publishing a small newspaper. Though four years apart in age, Wilbur and Orville were inseparable; their interests and talents were in harmony; it almost seemed that they shared one brain between two bodies.

The brothers were among the first in Dayton to adopt a new craze — riding a bicycle with two equal-sized wheels, chain-driven. The hobby almost immediately progressed into repairing other folks' cycles. Soon they became manufacturers, selling their own design "Wright Special" at $18 including tires.

By remarkable coincidence, this step toward becoming entrepreneurs was taken just about when the 1896 "Aeronautical Journal" editorialized:

> "To learn to wheel one must learn to balance. To learn to fly
> one must learn to balance. Why not begin now?"

A further coincidence was that Otto Lilienthal's fatal last flight — reported in newspapers all over the world — took place that year. Unfazed by Lilienthal's death, the Wright Brothers devoted themselves to learning about flying. They read everything in the Dayton Library even remotely relevant to that enthusiasm — even "Animal Mechanism," with sequential photographs of birds in flight. This became a hinge-point in history.

> "We could not understand that there was anything about a
> bird that would enable it to fly that could not be built on a larger
> scale and used by man," Wilbur concluded. "If the bird's wings
> could sustain it in the air, we did not see why man could not be
> sustained by the same means."

They read Chanute's book. They read Langley's "Experiments in Aerodynamics" and his "Experiments in Mechanical Flight." They studied Lilienthal with great care, particularly his calculations relating wing expanse to wind strength and weight of craft. His emphasis on the pilot maintaining complete charge of his vehicle at all times had special relevance to bicycle-riders.

The brothers' next step was to learn directly from the teachers — the birds. Through field glasses, they studied by the hour. Watching buzzards, they acquired a control theory by noting how the vulture was able to circle slowly in the sky over food on the ground; the birds

> "adjusted the tips of ... (their) wings so as to present one tip
> at a positive angle and the other at a negative angle ... thus ...
> turning ... into an animated windmill, ... when its body had
> revolved ... as far as it wished, it reversed the process and started

turning the other way. The balance was controlled by utilizing dynamic reactions of the air instead of shifting weight."

This insight alone carried the Wrights well beyond Lilienthal. But it took weeks to adapt the buzzard observations to glider-wing design. First they tried levers and pulleys. To be strong enough in combating wind gusts, the lever and pulley hardware would be too heavy to manipulate. Finally, they modified the kite by providing hinges for its tips, controlled with cords operated from the ground. By pulling on one cord or the other, the kite could be coaxed into changing direction.

Painstakingly, step by step, in this manner they accumulated more theoretical knowledge than Lilienthal or any other predecessor had possessed — without yet testing theory in manned glider flight. Wilbur wrote to Octave Chanute — the first letter in what would become a long correspondence and a close friendship:

> "For some years I have been afflicted with the belief that flight is possible to man. My disease has increased in severity and I feel that it will soon cost me an increased amount of money if not my life."

In this breezy idiom, he later confided:

> "Lilienthal's apparatus ... (was) inadequate, not only from the fact that he failed, but (in failing to recognize) ... that birds use more positive and energetic methods of regaining equilibrium than that of shifting the center of gravity."

Meanwhile, Langley was making slow progress toward building a full-sized airplane — four times larger than the original model glider, flown in 1896. It took all that time — years — by a fulltime staff in a fully-equipped machine shop. A gasoline engine also had to be designed and built; also two propellers with transmission gears, drive-shafts, and collateral equipment to transform internal combustion into forward thrust. Langley engaged Frank Manly to build a fixed-radial engine rated as capable of generating 52 horsepower. It would lift the craft, Langley said, skyward at a speed of 60 miles per hour. The pilot's "cockpit" was positioned under the fuselage which was festooned with balloon-like floats so that the 'plane could come down in water.

Langley had three years of headstart over the Wright Brothers. While Langley and his small army were building their full-sized airplane, the Wright Brothers' experiments with an unpowered glider were just starting. Wilbur Wright reached Kitty Hawk September 13, 1900. Neither brother had been there before. The site was chosen by correspondence with the U.S. Weather Bureau, followed by contact with the Coast Guard weather station. Between them, it was learned that reliable winds blew in the fall; the tallest obstruction was a bare hill 80 feet high; and there was ample space; almost complete solitude. Kitty Hawk was a tiny settlement of houses — less than two dozen — and a post office, on a barrier-island with empty beaches up to a mile wide and five miles long, almost completely barren of trees and bushes.

Wilbur's voyage from the mainland had been over choppy waters in an ancient fishing boat. But carefully-wrapped parts of their test glider were dry. Everything had been pre-cut in Dayton: ribs, spars, and wing sections. It took Wilbur two weeks to fit the glider together; he boarded meanwhile with a local family a half-mile walk from the work-site he'd chosen. When Orville arrived, September 29, they erected a 12x22-foot tent, anchored to one of the few trees sturdy enough to survive the notoriously stormy winters. It would house cots and cooking facilities, in addition to work-space.

In two days, the brothers assembled the craft. Influenced by Chanute's biplane design, it consisted of two wings five feet wide, cambered so as to conform with Lilienthal's lift tables. They were joined by five-foot struts and braced by Chanute-designed trusses. The wings were covered by lightweight sateen — except for an 18" area in the center, where the pilot would lie prone, strapped face-down. (Imagine the nerve! The Wright Brothers were willing to face — literally to face — coming in for a landing at 30 mph with head and eyes only a couple of inches above the ground!) The pilot's hands controlled the lines to the hinged air-foils on facing edges and trailing edges of the wing. At the very front of the glider, a horizontal rudder was affixed; by pulling on the lines this fragile accessory could control ascent and descent of the glider. The pilot's feet were fastened to a bar which was connected by lines to the wing tips; foot movements would thus maintain balance in flight. There was no fuselage. The glider was, in essence, a giant kite — about three times larger than the one they had experimented with, back in Dayton.

To become familiar with the larger craft, they first flew it unmanned, loaded with the deadweight of a bag of chains on the "pilot space" at the center of the lower wing. On very gusty days, the brothers learned the techniques of controlling the glider, as they had when boys, with a kite; acquiring knowledge that nobody before them possessed.

One day, after they had landed the unmanned craft to make adjustments in the control mechanism, a gust of wind lifted it off the ground and then hurled it to earth at some distance. Damage to the glider was extensive and took days to remedy. Setbacks like this required many hours of tedious, tiresome repair work. Experimental flights also revealed structural flaws. They learned from each failure. Several times, one of the brothers had to take parts back to their bicycle shop — four days round-trip — to be repaired, rebuilt, re-machined or replaced. Much time was lost in this manner. The season advanced, days shortened, wind grew stronger. The tour of duty at Kitty Hawk ended October 23, 1900. The brothers returned to Dayton, confident that they were making progress.

In July, 1901, they hired their first — and only — employee, Charlie Taylor, to take care of the bicycle-repair business. At Kitty Hawk, they pitched camp miles south of their original site, at the base of the high hill which would both shelter the shed they now set about building — and provide a downhill slope for a runway from which to launch the new, much larger, glider. The shed was to be large enough to house the glider, so work could be continued even when daylight faded. As before, all the glider parts had been cut and fabricated to size, to be assembled on site. The new glider was larger than the first, with wing expanse — almost double — 308 square feet. The craft weighed nearly 100 pounds; the brothers needed helpers from the Coast Guard station to drag it up the hill.

The first attempt at manned flight failed when the glider plopped to the ground immediately after being airborne. The "pilot-position" was shifted rearward; again the test failed. Flight by flight, inch by inch, the brothers found exactly the fulcrum location from which the pilot should operate the trailing edges of the wings. With many such slight adjustments, the craft finally could be "piloted" into rising as high as 400 feet at nearly 30 miles per hour.

The brothers also learned more about how to counter abrupt nose-dives and nose-up episodes, which Lilienthal had failed to deal with. More than once, they almost went into a stall; at times rate of climb was so abrupt as to invite stall. They reduced wing-camber to modify sensitivity to sudden wind-drafts. Even so, the glider came to earth frequently with severe impacts; on one occasion, the wings were so damaged that the fabric coverings had to be replaced in patchwork sections, sewn tightly by the postmaster's wife. The work went on day and night; even major repairs required less than a week.

The problem that continued to baffle Wilbur and Orville — tendency of the glider, without warning, to go into a spin out of control — did not yield to skill at the controls during flight. The problem persisted, unsolved through to the end of the season, nearly September. Chanute sometimes visited for weeks at a time; more than once with an entourage of glider enthusiasts.

Back in Dayton, the brothers checked and re-checked each and every Lilienthal calculation against every corresponding measurement of their glider. The suspicion grew that the German's tables might be inaccurate in charting the lifting effect of air-pressure on wing surfaces during flight. But there was no way to confirm that suspicion without actually building a series of gliders with different-cambered wings. Or was there?

The brothers devised a primitive testing machine, by cannibalizing several bicycles. From one, they removed handlebars. From another, they took a wheel and affixed it horizontally to the upright frame from which the handlebars had been removed. By exchange gear, the horizontal wheel revolved as the pedals drove the first cycle's vertical rear (drive) wheel, raised off the ground by a stand. To the horizontal "test wheel" two small plates were affixed. One was angled in fixed position as the Lilienthal tables mandated; the other was adjustable to reflect the actual speeds at which the wheel revolved. From the findings, they provided their own tables, amending those inherited from Lilienthal.

What should the correct air-pressure statistics embrace at a wide range of speeds? Wind-tunnels had been known for a quarter-century — but not like that the brothers then built in their second-floor workshop. It was a 6-foot-long oblong box, 16" square, open at one end and equipped with a fan. A window on top enabled the brothers to observe the behavior of the "airfoils." One was "fixed" — representing the Lilienthal angle. The second was movable, could be tilted at varying angles and speeds. Wind was created by the fan, in turn driven by a small engine. An air-pressure gauge reported performance at every angle and variable speeds. In essence, they observed air-dynamics of 48 different "wing" cambers! Weeks of close observation produced statistics that are still respected today as nearly perfect. "Breakthrough" is a tame word to describe the audacity and success of this crude device. Orville later commented: "I believe we possessed more data on cambered surfaces, a hundred times over, than all of our predecessors put together."

It was not an easy-won victory. More than 10 years later, in "Flying" magazine, Orville confessed

> "With the machine moving forward, the air flying backward, the propellers turning sidewise and nothing standing still, it seemed impossible to find a starting point from which to trace the various simultaneous reactions. Contemplation of it was confusing. After long arguments we often found ourselves in the ludicrous position of each having been converted to the other's side, with no more agreement than when the discussion began."

Their assistant, Charlie Taylor, remembered that "Both boys had tempers. They would shout at each other something terrible. I don't think they really got mad, but they sure got awfully hot!"

They reported each stage of progress to Chanute who publicized their findings; the Wright reputation spread. Wilbur was persuaded to speak on the "state-of-the-art" of manned, powered flying at a Chicago meeting of the Western Society of Engineers June 18. He caused a near-sensation when calmly proposing that certain of Lilienthal's calculations were incorrect. The rest of the summer was devoted to applying their wind-tunnel findings to revising wing-camber and wing-tip operation.

August 28, 1902, they returned to Kitty Hawk and spent 10 days repairing and enlarging the workshed which had, in its exposed location, been heavily damaged by winter storms. The sleep-area was moved into a loft to provide more floor space for work. Then they assembled the new glider. Wingspan had been increased one-third; the wings themselves made narrower; wing camber flattened. Though the sheet-metal forward-edge airfoils were still hand-operated, fractional movements of the pilot's hips were now sufficient to operate wingtips up and down. Rudders were added in the hope they would end the spin problem. Orville noted in his diary: "We are convinced that the trouble with the 1901 machine ... (will be) overcome by the vertical tail."

But he was wrong. Squirming his hips to operate the wingtips didn't prevent Wilbur, on September 23, from slipping into an uncontrollable dive, like Lilienthal's, and crashing to the ground. Luck held; he wasn't injured. Three days were needed for repairs. Then every detail of flight control was double-checked. Orville finally reasoned that the new rudder component might be the problem. Modification of the tail assembly was undertaken, making it movable. Movable by what? The pilot's hands and feet already were fully engaged in operating the ascent-descent elevator; his hips were simultaneously in motion, linked to working the wing-warping wires. Wilbur designed a chest-harness with which he could operate the rudder, shrugging the control-line even while his feet operated the wingtips.

From then on, it was clear sailing — and clear gliding. The brothers competed in the length and speed of their test flights, reaching over 600 feet in less than half a minute. They soared joyously over dunes and water, gliding, floating up and down, banking, landing gently at will. They returned to Dayton at the end of the flying season, in high spirits, anticipating the next series of tests which would take them aloft with an engine.

Through the winter, Langley — now aware that he was in a race, rushed toward his final test. Smithsonian employees went on overtime. His brilliant engine designer, Charles Manly, also had been working day and night for four years, to create a water-cooled 200-pound 52-horsepower engine with carburetor, and a train of gears and shafts to connect with the pusher-propellers.

This would also be a Potomac launch, at Wide Water, Virginia, where a houseboat had been towed by two tugboats from Washington to its anchor-position. Atop the "house" portion of the boat a 60-foot catapult rail had been bolted. On it, the airplane was tethered to a restraining spring. Manly wormed his way through struts and guy-wires into the "cockpit." He revved the engine and waved the order to release the catapult brake. The craft shot forward along the launch rail, out beyond the stern of the 30-foot houseboat, and ...

Dropped, like a stone, into the water. The mechanism to launch the 'plane had caught on the rail at its end, and pitched the craft nose down into the water. Manly was fished out from under wreckage.

Langley and his crew returned to the Smithsonian for more adjustments to the craft and to the rail. The next test was ready by the late afternoon December

8, 1903. With light failing and the temperature dropping, Manly — this time in a cork-lined jacket over long underwear — again crawled through the bracing wires, into his cockpit. He signaled to release the catapult-spring, shot 60 feet along the catapult rail, rose into the air, and ...

Flipped upside down before settling into the water. Manly found himself under an ice-floe and had to swim some distance under the surface to escape. Luckily he was a good swimmer, and returned to the boat while rescuers in rowboats frantically searched the wreckage for him.

The government's $50,000 had by now been spent to the last penny. Langley, at 69, had the good sense to accept the cruel truth that he did not have the tenacity — not to mention money — to start over yet again, to continue gambling with the mysteries of aerodynamics and the realities of risking a man's life. He dropped out of the race.

His decision was of course not known to the Wright Brothers. During the winter they had designed and built a 140-pound engine that generated 12 horsepower. Then it took more than two months to glue together layers of spruce wood — and whittle them into a pair of 100-inch propellers. They were shaped differently from any propeller known before. As always, their work was guided by inspired improvisation. Their employee, Charlie Taylor, recalled:

> "We didn't make any drawings. One of us would sketch out
> the part we were talking about on a piece of scratch paper and
> I'd spike the sketch over my (work) bench."

Orville's exuberant words in a letter to a friend: "we worked out a theory of our own ... and soon discovered, as we usually do, that all the propellers built heretofore are all wrong!"

They reached Kitty Hawk September 25, 1903. As usual, they had to rebuild the largely destroyed shed, torn by hurricane winds from its foundation. The campsite was in a lake; it was surrounded by waters left from torrential rains. In the stagnant pools, millions of mosquitoes had bred, and the Wrights were plagued at all times by swarms of the voracious insects. Orville wrote home:

> "The rain has descended in such torrents as to make 'a lake'
> for miles around our tent; the mosquitoes were so thick that they
> turned day into night ... the lightning so terrible that it turned
> night into day."

While they waited for arrival of the latest model of glider — shipped as its predecessors had been, pre-cut, each part separately wrapped — they practiced in the earlier model. By tinkering with it, they made it ever more responsive to control, and gradually reached a point where the pilot could coax it into hovering in the sky, almost motionless — like their old teachers, the buzzards. One flight of about 50 feet had the glider aloft, almost still in the sky, nearly half a minute. Mastery of the air was now measured in time aloft, not distance flown.

A two-day gale that whipped winds up to 75 miles per hour tore away part of their shed roof. Albemarle Bay tidewater surged into the campsite, flooding the work-floor. The men doggedly made repairs in a few days.

When the newest-designed model glider arrived, it was hastily assembled. It was bigger and heavier than any of those they had flown before. Each wing consisted of three sections. The two at the ends were warped, as before; but the center section, supporting pilot, engine and drive-shaft, was rigidly trussed. The wings were sheathed in undoped muslin. Total weight was considerably more than the Wright Brothers would have liked, but now they were so confident they knew what they were doing, they were sure that their engine, rated at better than eight horsepower, could lift and keep it in the air.

The first test, however, resulted in yet another setback: both propellers tore loose and their shafts were damaged. It required 10 days for repairs, in Dayton. Back in Kitty Hawk, trusses were tightened; the launch-rail from which the powered glider would take off was inspected inch by inch. The weather grew so cold that their hands froze and work had to be suspended. Patches of ice appeared wherever pools of water collected.

Friday, November 20, the repaired propeller shafts arrived and the blades were fitted. With engine purring, however, they did not rotate. The sprockets had worn down in the previous test, and were loose. The Wrights luckily had brought with them heavy-duty adhesive cement that normally was used on bicycle tires. They liberally "layered" it onto the sprockets. Orville's diary note: "we stuck those sprockets so tight I doubt that they will ever come loose again."

On the 28th of November, all was ready for the powered flight. Disappointment again: repeated "dry runs" revealed a flaw in one of the repaired propeller shafts — a hairline crack. Orville rushed back to Dayton on November 30, in order to replace the wood housings with solid steel. He returned to Kitty Hawk December 11, after a two-day journey. The Wright Brothers still had no idea of Langley's disappointments.

The repaired part was installed December 12, but there wasn't enough wind for glider operation. The day after was better, but considering the weight of the new machine, the brothers decided they would need the assistance of gravity to generate takeoff speed. So the craft was carried up the hill. From the summit, they lay a 60-foot track at an angle just under 10 degrees. The brothers flipped a coin to decide who would be first. Wilbur won. The two shook hands. He climbed in. Orville positioned himself at wingtip; a friend from the Coast Guard station took the other tip, to steady the machine as it rolled down the monorail. Failure again, as the plane gathered speed greater than the men could run - and foundered. Another day was needed, back in the shop, to make repairs.

There followed two days of insufficient wind, during which the brothers agreed to abandon the downhill-momentum idea for takeoff. At dawn December 17, finally, a freezing wind blew out of the north — over 20 mph, and getting stronger by the minute. The brothers shaved, dressed carefully, complete with white shirts, clean celluloid collars — and neckties! By mid-morning, all was ready for a launch. They walked to the rear of the airplane, spun the propellers into life, shook hands.

A quarter-century later, one of the Coast Guard Station men recalled in "Collier's" magazine that

> "we couldn't help (but) notice how they held on to each other's hand, sort o' like two folks parting who weren't sure they'd ever see one another again."

Orville climbed in. Wilbur lifted the right wing from the bench on which its tip had been resting. The restraining ropes were released, the plane took off into the north wind.

The next fateful minutes, the next fateful hours, are so moving in Tom Crouch's very readable biography, "The Bishop's Boys," that no better description can be imagined:

> "The airplane floundered forward, rising and falling for 12 seconds until it struck the sand only 120 feet from the point at which it had left the rail .... For the first time in history, an airplane had taken off, moved forward under its own power, and landed at a point at least as high as that from which it had started — all under the complete control of the pilot. On this isolated, windswept beach, a man had flown."

After a pause, Wilbur took his position stretched out on the lower wing. Into the air he went and flew 195 feet. Then Orville flew 200 feet in 15 seconds. Then Wilbur again, 852 feet in just under one minute. They stopped for lunch and planned longer flights for the afternoon, maybe all the way to the weather station. As next described by biographer Crouch:

> "Suddenly, a gust of wind raised one wingtip high into the air.... The engine broke loose as the disintegrating machine rolled over backward ... snapping wires and splintering wood ... the world's first airplane was transformed into a twisted mass of wreckage."

Unlike earlier models that had been abandoned to the ravages of wind and weather at Kitty Hawk, the Wright Brothers packed up the pieces of the 1903 airplane, and shipped it back to Dayton.

There is more to their story, of course. How they continued their experiments — 1400 flights! — back in Ohio; how they continued to improve their technology and techniques; their earnest and long effort to interest the American government in acquiring exclusive rights to their patents; why they failed and had to deal with foreign powers – always eager to acquire a better weapon for use in the regular conflicts with neighbors; how they had to spend years and a fortune to defend their patents against rapacious interests — large ones — seeking to infringe upon them; when and where they participated in air-shows. What has been synopsized in the preceeding pages here occupies but half of Tom Crouch's splendid book.

We will close our account by quoting the plaque on an airplane so small it looks like a toy hanging in the North Hall of the Smithsonian's Arts and Industries Building:

BY ORIGINAL SCIENTIFIC RESEARCH THE WRIGHT BROTHERS
DISCOVERED THE PRINCIPLES OF HUMAN FLIGHT
AS INVENTORS, BUILDERS AND FLYERS THEY
FURTHER DEVELOPED THE AEROPLANE,
TAUGHT MAN TO FLY, AND OPENED
THE ERA OF AVIATION